See Sooner, Act Faster

Management on the Cutting Edge Series from MIT Sloan Management Review

Edited by Paul Michelman

Published in cooperation with *MIT Sloan Management Review*

The AI Advantage: How to Put the Artificial Intelligence Revolution to Work
Thomas H. Davenport

The Technology Fallacy: How People Are the Real Key to Digital Transformation
Gerald C. Kane, Anh Nguyen Phillips, Jonathan Copulsky, and Garth Andrus

Designed for Digital: How to Architect Your Business for Sustained Success
Jeanne W. Ross, Cynthia Beath, and Martin Mocker

See Sooner, Act Faster: How Vigilant Leaders Thrive in an Era of Digital Turbulence
George S. Day and Paul J. H. Schoemaker

See Sooner, Act Faster

How Vigilant Leaders Thrive in an Era of Digital Turbulence

George S. Day and Paul J. H. Schoemaker

The MIT Press
Cambridge, Massachusetts
London, England

This book was set in Stone Serif and Stone Sans by Westchester Publishing Services. Printed and bound in the United States of America.

Library of Congress Cataloging-in-Publication Data

Names: Day, George S., author. | Schoemaker, Paul J. H., author
Title: See sooner, act faster : how vigilant leaders thrive in an era of
  digital turbulence / George S. Day and Paul J. H. Schoemaker.
Description: Cambridge, MA : MIT Press, [2019] | Series: Management on the
  cutting edge | Includes bibliographical references and index.
Identifiers: LCCN 2019006921 | ISBN 9780262043311 (hardcover : alk. paper)
Subjects: LCSH: Leadership. | Management--Technological innovations. |
  Vigilance (Psychology)
Classification: LCC HD57.7 .D7347 2019 | DDC 658.4/092--dc23
  LC record available at https://lccn.loc.gov/2019006921

10 9 8 7 6 5 4 3 2 1

Contents

Series Foreword

The world does not lack for management ideas. Thousands of researchers, practitioners, and other experts produce tens of thousands of articles, books, papers, posts, and podcasts each year. But only a scant few promise to truly move the needle on practice, and fewer still dare to reach into the future of what management will become. It is this rare breed of idea—meaningful to practice, grounded in evidence, and *built for the future*—that we seek to present in this series.

Paul Michelman
Editor in chief
MIT Sloan Management Review

Acknowledgments

A handy metaphor for conceiving and writing a book is taking an ancient caravan journey. It starts with a vision of the destination and enthusiasm about the possibilities. While underway, it requires persistence and patience to deal with the inevitable delays and detours. Most of all, the writing journey draws on the support of colleagues, friends, and family members to reach its destination.

We came together on this journey because of mutual respect and a history of successful collaboration. We maintained our compass heading with the help of an extended community of thought leaders, scholars, and business leaders. Our colleagues at the Wharton School and especially at the Mack Institute for Innovation Management were invaluable at many stages of the journey. We also benefited from fruitful personal interactions with professors Barb Mellers, Saikat Chaudhuri, Tom Donaldson, David Reibstein, Harbir Singh, J. Edward Russo, Greg Shea, David Teece, Michael Useem, Phil Tetlock, and Jerry Wind.

During our journey of discovery and learning, we benefited from clients such as Abbott Labs, Novartis, Cancer Treatment Centers of America, Ferrero, Knight Foundation, European Central Bank, Medtronic, the Prime Minister's Office of the United Arab Emirates, UBS, and W. L. Gore and Associates. These and many others allowed us to apply our ideas and methods to their adaptation challenges, yielding invaluable feedback. Especially useful guidance came from participants in executive education programs where our ideas often were first discussed. We have also both benefited from our experiences in starting companies and serving on diverse boards of directors. These real-world settings deepened our appreciation of the key roles that leadership teams play in guiding and completing any journey.

Many books are started, but few successfully reach their final destination. Karen Christensen and Kirsten Sandberg served as our editorial "harbor pilots," taking the reader's perspective to challenge us to develop persuasive and engaging narratives. We are grateful to all those who gave us constructive feedback along the way, including Charles Baden-Fuller of the University of London, Kaihan Krippendorff (founder of Outthinker Roundtable), Doug Randall (founder of Monitor 360), Govi Rao (former CEO of Noveda Technologies), Paul S. Schoemaker (President of Public Salt), Michael Taylor (CEO of SchellingPoint), Alan Todd (CEO of CorpU), and Toomas Truumees (partner at Heidrick and Struggles). They gave freely of their time to discuss nuances in the chapters while pressing us to further clarify our concepts and explain the reasons behind our recommendations. Along the journey, Erika Burnett at Wharton kept the manuscript moving forward with skill and unfailing good cheer.

Just as maps and satellite navigation systems are essential to travelers, so were the in-depth surveys we collected from 335 vigilant and vulnerable organizations to validate our model. Senior participants in many executive programs, leadership meetings, and conferences were a great help as well. Special thanks go to the organizations that saw the possibilities of our journey and gave us high-level access to their members. Mette Laursen of LinKS in Denmark, Sam Gill of the Knight Foundation, Vanita Bhargava of the Council on Foundations, and George Hofheimer of the Filene Research Institute, all went above and beyond to adapt our survey and the findings to their own distinct environments.

A writing journey needs sponsorship, encouragement, and hard-nosed feedback from a publisher. The editors at the MIT Press were early to see the potential of our book project, especially why digital turbulence increasingly poses new challenges for leaders and organizations. We benefited from the broad vision of Paul Michelman (editor in chief of *MIT Sloan Management Review*) and the sure-handed editorial oversight of Emily Taber (acquisition editor at the MIT Press). They and their staffs were a pleasure to work with and very helpful along the way.

Above all, we are grateful for the understanding and unconditional support of our dear wives, Alice and Joyce, who sustained us at every step of the writing journey. We dedicate this book to them.

George S. Day
(Villanova, Pennsylvania)

Paul J. H. Schoemaker
(Delray Beach, Florida)

Introduction

> Turbulence, by definition, is irregular, non-linear, erratic. But its underlying causes can be analyzed, predicted, managed.
>
> —Peter F. Drucker[1]

We live in an increasingly turbulent world, filled with both leadership dilemmas and unlimited opportunities. This relentless turbulence can be managed, just as whitewater rapids are navigable with a vigilant guide. But it demands different capabilities from those used to manage the current operations. This book is for leadership teams that need their organizations to become more vigilant and able to navigate the increasing turbulence magnified by the inexorable process of moving from analog to digital.

Vigilant firms have greater foresight than their rivals. Charles Schwab was early to see and act on the promise of "robo-advisors," GM jumped ahead of Ford in the world of autonomous cars, and Novartis led others in equipping their sales teams with a digital platform to access medical expertise for doctors in real time.[2] Such vigilant firms win at the expense of their slower rivals. Vulnerable firms, on the other hand, often misread early signals of external threats or internal organizational challenges. Honeywell stumbled when Nest Labs came out first with a sleek, Internet-enabled thermostat, an early tool in the creation of the nascent "smart home." Volkswagen crippled itself with revelations that it created intelligent systems inside its cars to fake emissions test results. Dansk Bank, Denmark's largest lender, spiraled down from being one of Europe's most respected banks after being caught in a massive, 200-billion-euro money-laundering scandal in which it ignored many red flags.

The myriad problems afflicting Facebook that first surfaced in 2018 show that vigilance is about how organizations frame or ignore problems, interpret awkward information, share information, and deal with difficult questions—or not. The infamous "deny, deflect, and delay" approach of CEO Mark Zuckerberg and COO Sheryl Sandberg to dealing with manipulative and false campaign messages, illicit data sharing via Cambridge Analytica, and hate speech has had severe effects on the firm. Facebook squandered some of the essential trust on which its social media platform was built, unleashed angry calls for regulation, and bashed both its reputation and market capitalization (within one five-month period, the stock price dropped 39 percent). This issue goes well beyond the Facebook malfeasance and is a warning sign to all organizations that their ownership and monetization of the individual data they hold is becoming tenuous.

The antidote to such vulnerabilities is *heightened vigilance*. Traditional methods of strategic planning, risk analysis, and decision modeling are now less effective because there is just too much uncertainty on the periphery and too little stability at the core. Today's environment requires new tools and mindsets. Fortunately, our understanding of the nature of vigilance—including corporate foresight—has advanced steadily. The fulcrum of strategic thinking has moved from leveraging a firm's scarce and valuable resources to building the dynamic capabilities[3] needed to adapt to increased turbulence. Whereas the resource-based approach emphasizes internal efficiency and narrows the strategy dialogue, a dynamic capabilities approach puts the emphasis on organizational anticipation and agility—enabling a firm to shape the environment to its advantage.

Vigilance is rewarded when there is a corresponding ability to act faster than rivals once the ambiguities of a weak signal of a threat or opportunity are resolved. In this realm, we'll employ the latest insights about design thinking, including "fail-fast" experiments and investing flexibly using "strategic options." Such options are small enough bets that a firm can unwind them if necessary while still learning enough to stake out market positions or build new capabilities. Creating an "ambidextrous" organization that supports both innovation and the optimization of current models is another crucial ingredient.

Vigilance means sensing, probing, and interpreting weak signals from both inside and outside the organization. Highly publicized internal breaches such as the Wells Fargo account manipulation scandal and the revelation that numerous Japanese firms didn't follow their quality standards[4] have created shock waves in boardrooms. These firms and many others have been damaged by scandals that could—and should—have been foreseen. All have paid a high price, with fines and legal costs being just the tip of the iceberg. The larger costs come from managers being distracted by the crisis, as well as collateral damage to a brand, the morale of a company's employees, the hesitation of new partners and customers, and the burden of greater regulatory oversight.

One certainty about vigilance is the pivotal role of the leadership team in preparing and aligning the organization to see sooner and act faster. Modeling such vigilance is a collective skill set most evident in its absence. The words no board or investor ever wants to hear about an organization's leaders are "they ignored the warning signs" or "they missed the boat." Boards don't expect prescience, but they do rightly rely on the leadership team to see and act on early warning signs of trouble or opportunity. Through a diagnostic survey that we use during our client engagements, we've learned a great deal about what sets a vigilant leadership team apart. This diagnostic tool is in appendix A of this book. We use the data collected from 118 firms around the globe to test and validate our approach to building a vigilant organization.

Looking Forward

Attention is one of the scarcest resources of any organization and is quickly depleted in a digital environment spewing out huge amounts of background noise. Yet these same digital technologies can also help us to manage information overload by separating signal from noise and steering attention to where it's most needed. Data mining, predictive analytics, knowledge management, scenario planning, and neural nets, to name a few, help to filter signals and highlight those posing a real or imminent risk. Organizations can likewise leverage crowdsourcing technologies, open innovation platforms, and tighter links with network partners to identify new customer requirements and develop products to serve them.[5]

Responding to a bewildering array of ambiguous signals and sorting them from the surrounding noise requires a judicious balancing of speed and cautious exploration. Although these attributes are not the usual hallmark of large organizations, some vigilant firms have learned to be nimble and act faster than their rivals—building a stronger, more agile organization and an enhanced reputation for foresight and innovation, which in turn yield greater customer and shareholder value.

In chapter 1, we explain why there is less stability within the core of most organizations today—and greater uncertainty on the periphery. We share hard-earned lessons from our own experience and that of others as the basis for our approach to becoming vigilant. A pivotal lesson here is that being able to see sooner and act faster preserves more strategic degrees of freedom for leaders. When the digital fast-forward button has been pushed, few organizations can afford to wait to be sure they know exactly what is happening and then react. "Waiting for the fog to lift" typically produces inaction and eventually leaves an organization with few viable ways to respond. The graveyard of business history contains thousands of examples of slow and misguided actions from overly conservative leaders.

Chapter 2 describes the four defining features of vigilant organizations: the stance of their leadership, how they approach strategy making, invest in foresight, and then ensure coordination and accountability. We use Adobe Inc.'s revitalization strategy as an instructive example of how vigilant leaders can venture into the unknown before digital shockwaves knock over the furniture. This chapter also shares some of the findings from our study of 118 diverse organizations and discusses how our survey in appendix A can be used as a diagnostic tool to focus a leadership team on how best to sharpen its vigilance. The postures needed for success in digital markets are often different from traditional management approaches. They are more lateral than vertical and more exploratory than determinative. There is a premium for creativity when navigating new terrains with low visibility and unexpected undercurrents.

Whereas chapters 1 and 2 set the digital stage and lay out the challenges at hand, the remaining chapters focus on what leaders can and should do. The flow of chapters 3–6, shown in figure 0.1, follows the contours of an organization's choices about where to allocate their scarce attention resources, how to sense weak signals, sort them from background noise, and then act on those signals before rivals respond. We offer tested guidance for setting up safeguards against biased interpretations. Chapter 6 outlines the approaches leadership teams can use to act quickly in response to signals deserving attention. The challenge is not just fast action but *wise action* that includes knowing when to be patient and "how much powder to keep dry."

Chapter 7 presents an action agenda for leadership teams to follow to enhance strategic vigilance throughout their organizations. We connect *seeing sooner* and *acting faster* to the capabilities that provide an organizational gyroscope to keep the firm balanced in a turbulent world. The conclusion (chapter 8) extracts the most important lessons from this book. These are the lessons all leaders can apply confidently while navigating digital turbulence.

This book aims to help leadership teams navigate deep uncertainty in general and *digital turbulence* in particular. Our guidance is grounded

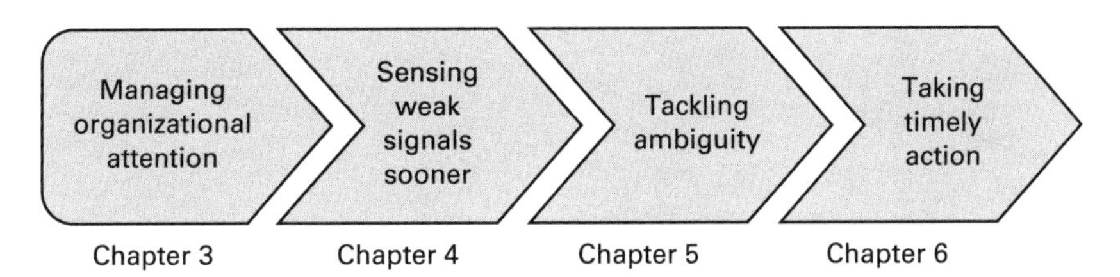

Figure 0.1
Flow of core chapters

in extensive, hands-on experience, which cannot readily be distilled to a simple, linear formula. This guidance entails a balance of art and science, with a premium for leaders who can blend both. Solving the challenges they face in trying to thrive amid turbulence takes patience, commitments of resources, seasoned judgment to ask the right questions, and timely action on the answers.

The vigilant leader's task is to master the art of anticipation of potential threats and opportunities, informed by hypotheses, facts, data, experience, and expert guidance. We encourage leadership teams to nurture superior vigilance capabilities to anticipate the challenges of digital turbulence and turn them into advantages.

1 Facing Reality in Real Time

The beginning of all wisdom is recognition of facts.

—J. K. Paasikivi, former president of Finland[1]

At some point in time, nearly every leadership team will miss a critical signal that could—and should—have been caught. In past eras, the slower pace of these missed threats might have allowed reactive organizations time to regroup and respond. But today's environment of digital turbulence—with its defining features of increased speed and transformative business models—increasingly penalizes tardy responses. By the time a digitally enabled competitor comes into full form, it may have already amassed insurmountable market power. Likewise, internal threats can grow and metastasize faster until they can't be contained.

What is needed is greater *vigilance*. Vigilant organizations see risks and opportunities sooner and position themselves to act faster to address them. Vigilance is much more than a single individual's heightened sense of alertness. It is a collective capability that firms must nurture, characterized by curiosity, candor, and a concern for the long-run welfare of the organization. Above all, vigilance is a superior ability to anticipate serious threats, recognize major opportunities, and then act faster than others despite incomplete knowledge.[2] Vigilant organizations gain flexibility and time by testing prototypes, taking small bets early on market experiments, and making exploratory acquisitions. These strategic options are easy to unwind at any time but give the firm a head start when the fog of uncertainty eventually lifts. Without this flexibility, firms have to react to fast-moving events and lose their degrees of freedom to maneuver.

Unfortunately, *vulnerability*, not vigilance, is the norm in most companies. The immediate pressures of running day-to-day operations consume so much of the scarce attention of the leadership team that it lacks the time and resources to look beyond immediate concerns. The result is missed signals, like the ones that led RadioShack in the mid-2000s to devote huge proportions of its limited space and staff across its seven thousand retail stores to serving the growing population of confused first-time smartphone users—a segment quite different from its loyal electronic hobbyist clientele. The company didn't foresee that wireless carriers would start to compete seriously by opening their own retail stores. This put RadioShack in head-to-head competition with powerful corporations like AT&T and Apple while alienating its hobbyist base. When the next do-it-yourself intelligent electronics fads hit the scene—3-D printing, personal drones, and smart home kits[3]—hobbyists simply went elsewhere. RadioShack was deemed no longer relevant.

Another costly example of poor vigilance is the long list of major hotel chains that missed the consumer trends that paved the way for Airbnb. Founded in 2008, this peer-to-peer property rental platform brilliantly tapped into the unmet demand for convenient, comfortable, and affordable accommodations in popular travel destinations. Missing warning signs is not a failing limited to the weak or the poorly led. Even the leadership teams of respected and well-run organizations can be blindsided.

We have followed Mattel since 2006, when the Barbie doll was deposed as queen of the fashion dolls by a brash upstart doll brand aptly named Bratz.[4] This challenger targeted precocious older girls by depicting pouty-lipped teenagers with attitude and flair, whereas Barbie—with her fairytale outfits—appealed to the childhood princess fantasy popular with three- to five-year-olds. The defection of older girls was very costly to Mattel, since the toy maker relied on Barbie for 30–40 percent of its operating profits. When Mattel belatedly woke up to the Bratz threat, it reacted with a lame, copycat doll called Flavas.

Mattel's original myopia was caused by a narrowing of leaders' collective attention due to an internal cash flow crisis. An unfortunate acquisition had led to a huge loss, and new leadership came from outside the toy industry to turn the company around. Although the ensuing cost cutting improved margins, sales of Barbie dolls continued to slide. Mattel's vulnerability was compounded by an inside-out, product-centered, and reactive culture, with a strategic focus on maintaining the Barbie franchise and brand.

Mattel continues to lag behind rivals who are bringing digital technologies to the realm of imaginative play, adding interactivity and customization. They have been slow to see the potential of augmented reality (AR), which takes interactivity to a new level by superimposing digital images on top of our view of real-world objects. This is different from virtual reality (VR), which fully replaces a physical reality with a computer-generated environment.[5]

Mattel had an early signal of AR's potential with the booming success of Pokemon GO, an app allowing users to hunt for Pokemon in the real world using an app on their mobile phones. Others saw these possibilities, and in October 2017, Parker—the first AR teddy bear—was launched by the tiny Seedling USA which was founded in New Zealand in 2007.[6] A smartphone app allows children to interact with the bear, learn about Parker's inner organs, and even diagnose illnesses. The future of toys will be digitally integrated and Barbie cannot afford to be late again.

How Digital Technologies Create Turbulence

Digital technologies are transforming how we process information, learn, make decisions, and interact with each other, and no industry is immune from the resulting turbulence. In 2005, Amazon was an online bookseller disrupting the physical and e-book markets. Today, it has much broader reach—captured in the melodramatically named Death by Amazon Index, which tracks the impact of the online giant on fifty-four retailers (and counting) that have been adversely affected.[7] The company's reach goes far beyond retailing. Amazon can use its expertise in data analytics and deep customer insights to move into almost any industry. As recently as 2005, Yellow Cab dominated personal transportation, and hotels offered the primary option for temporary housing. Now we have Uber, Lyft, and Airbnb, the business models of which leverage the widespread adoption of smartphones. These new entrants pursue asset-light strategies, transcend traditional vertical markets, use fast-scaling platform strategies, leverage social media, and can pivot quickly when necessary.

Multiple technological shifts have enabled these companies to reach market dominance in record time. Plummeting hardware and software costs have placed vast quantities of computing power into ever-smaller devices, from the mobile phone in your pocket to the Fitbit on your wrist with a glucose monitor embedded under your skin to the billions of sensors, cameras, and microphones creating, accessing, and processing signals (and noise) from every process imaginable. This exponential data explosion is such that various data experts estimate that 90 percent of all the data in history has been created in the last two years.[8]

If Gordon Moore's 1965 paper on computational trends[9] is the starting point, the digital, computational, and communications revolutions have been underway for more than fifty years, and we are now seeing the consequences of a billionfold improvement in performance. Coming over the horizon are equally dramatic improvements in digital fabrication capabilities.[10] Today's 3-D printers presage a powerful trend to turn data into any kind of object, with applications for making everything from food to furniture. The possibility of hyperlocalized production of (almost) anything may one day overcome the constraints of long, global supply chains.

The interwoven nature of digital technologies is suggestively depicted in figure 1.1 and quickly brings to mind innumerable possible combinations of digital capabilities. This "ball of yarn" schematic is dauntingly complex, showing eight digital technologies around the outer ring. Each is both a *source* and a *result* of other digital advances, leading to yet more new capabilities.

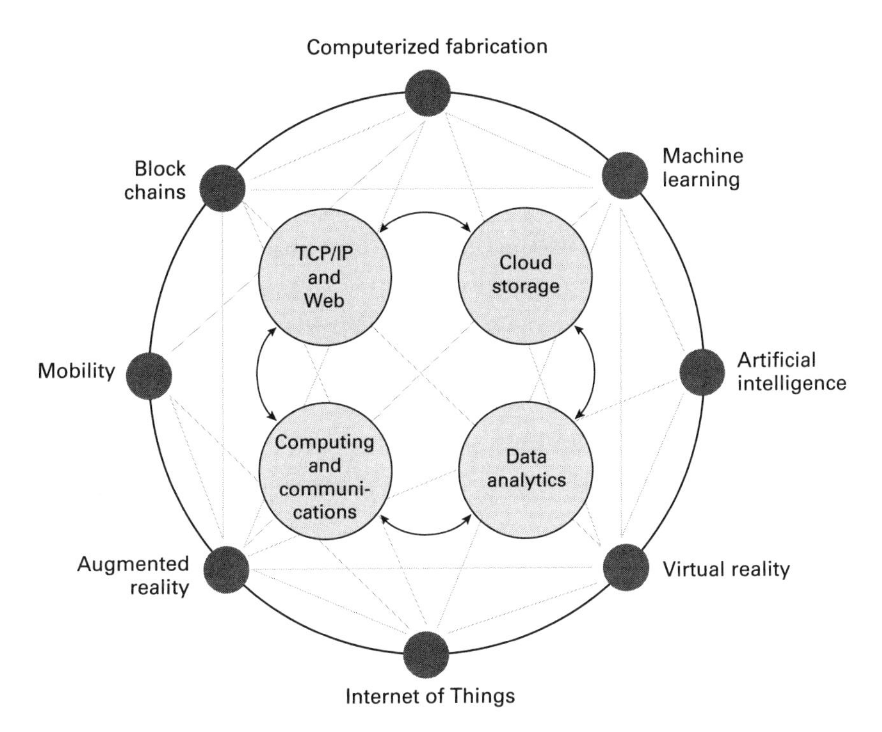

Figure 1.1
Combinations of digital domains

These digital capabilities are a product of breathtaking advances in computer system performance, including processing, storing, communicating, and analysis, as partly shown in the inner ring of figure 1.1. For example, artificial intelligence (comprising a set of "smart" technologies that can learn from their environments and take autonomous action) is enabled by advances in neural networks and silicon-level technology, aggregation of storage in massive data centers accessible via the cloud, and a host of other symbiotic advances.[11]

The combinations of digital technologies are typically complex and nonlinear, with often unanticipated interactions when they are applied. A technology that seems nonviable or commercially distant can suddenly take off when the stars align. For example, Honeywell was surprised when Nest Labs came out with a sleek, Internet-enabled thermostat, allowing harried commuters to activate their home systems remotely so that the lights would be on and the house heated when they arrived home. This convergence of technologies had been in the works for years, secretly at times or outside the purview of rivals. Although the incubation period of some digital technologies can be long, they can produce very fast technological change when all the pieces align. This seems to be happening with self-driving cars and electric vehicles now, although it's too soon to know for sure.

Our main point here is that no single digital technology enables digital turbulence. Rather, it is an effect of the simultaneous maturing of multiple parallel technologies, sharp declines in their costs, new functionalities, and the emergence of new platforms to put them to work. The unpredictability of these process, in terms of progress and impact, create turbulence.

Transience and Turbulence

An enduring belief of strategists is that sustainable advantages can't be purchased in open markets, but must be built internally through "rare, imperfectly imitable and non-substitutable resources."[12] Rita McGrath and others argue that even organically developed advantages, once believed to be defensible, permanent, and durable, may be transient as turbulence increases.[13] The implication she draws is that firms need a new set of dynamic capabilities for "surfing through waves of short-lived advantages." An apt metaphor she uses is that strategy making is becoming less like chess and more like the ancient Chinese board game Go. This complex game involves 150 moves on average, implying 10^{170} configurations. This nearly unfathomable degree of combinatorial complexity for the human mind puts a greater premium on lateral rather than linear thinking and computing. Not surprisingly, it took artificial intelligence a decade longer to beat the top human player in Go than in chess, with continued progress since.[14]

Most leadership teams realize that new linkages among these digital technologies can bring unexpected turbulence and disruption to their industries. Some are already tired of generic warnings of impending upheaval. They say, "Okay, we get the message, but what should we do about it?" They struggle to anticipate what may lie around the corner when certain situations will arise:

- Digital platforms help new global players to emerge in unexpected ways. China now has a large lead in the ability to make mobile payments (roughly fifty times that of the United States). In just fifteen years, the number of Chinese firms in the Fortune Global 500 has increased by more than twenty times.
- Market boundaries are blurring and dissolving. Financial technology, or fintech, is altering the nature of money itself, including how customers transact and secure loans. Big bets are being made by countries (e.g., Bermuda and Switzerland) and companies on blockchain technologies that enable cryptocurrencies for decentralized electronic exchanges of value, possibly hastening the obsolescence of cash.
- Complex ecosystems are emerging. This week's competitor may be next week's supplier, customer, partner, or all of the above. Although Apple and Samsung compete fiercely in the mobile phone market, Apple relies on Samsung for key components for its phones.
- The pace of change is accelerating. Time is being so compressed that the rate of change is exceeding the ability of traditional, hierarchical organizations to keep up.

Meanwhile, organizations are also grappling with ongoing changes in stakeholder and customer requirements, competitor strategies, access to resources, and the political and regulatory environment. Digital turbulence intensifies all these challenges.

The uncertainty created by digital technologies can obscure our vision of the future because it may rest on unexamined and misleading assumptions. For example, the rapid acceptance of social media platforms glossed conveniently over privacy concerns. Following the Cambridge Analytica scandal in 2018—when it surfaced that Facebook profiles were shared without user permission or knowledge—it dawned on people that their intimate digital data may be passed around in ways they never intended. As Sheryl Sandberg, Facebook's COO, later admitted, "We were too slow to spot this and too slow to act. That is on us."[15]

Another uncertain and dangerous assumption is that a digital technology advance will necessarily satisfy customers. Facebook, Google Search, and Intuit's Quicken all created customer value[16] by being easy to use, saving time, reducing risk, and improving productivity. But there may also be unintended consequences, as Google Glass experienced when it launched eyeglasses that display computer information. Google Glass allows users to communicate hands free with the Internet via natural language voice commands. Although it was a technological tour de force, it lacked any real style or fashion appeal and set off alarm bells related to privacy violations.

Increasingly, digital innovators confront direct systemic threats of hacking and cybersecurity breaches and indirect threats from system limitations. These uncertainties feed the whirls of turbulence. We can only suggest what to expect because digital technologies interact with each other, as well as with other trends, in unpredictable ways.

Bane or Boon to the Organization?

Due to their recombinant nature, digital technologies cut both ways. As a bane, leaders need to wade through a far larger volume of information—much of it noisy and irrelevant—to uncover relevant signals. As Nate Silver noted, "Information is no longer a scarce commodity ... but relatively little of it is useful, because useless data distracts us from the truth."[17] The cybersecurity breaches at Target, AOL, and credit-scoring firms were made worse because numerous alerts of system breaches were ignored until the hackers gained sufficient knowhow to compromise the entire system and capture sensitive customer data. The overload of false warnings eventually desensitized frontline computer analysts.

Advances in technology can also nurture internal problems and allow them to fester in the shadows. The scandal at Wells Fargo—in which customer service personnel created over two million unauthorized accounts for existing customers using digital means—is instructive. This misbehavior continued for years in the firm's retail banking and credit operations. The bank has paid $185 million in fines, put aside about $110 million to settle a class-action lawsuit, and fired 5,300 employees.[18] It appears that hundreds of thousands of Wells Fargo auto loan customers were also digitally registered for car insurance they neither needed nor approved. Sadly, tens of thousands had their cars repossessed for failing to pay for this insurance.[19]

The problem festered because bank systems allowed customers to open accounts without going to a branch or providing an ink signature. Tellers, customer agents, and even automated systems can create what appear to be legitimate accounts. They can also rig account preferences so that customers receive no statements and thus no signal that the account exists. As far back as 2005, Wells Fargo employees were blowing the whistle about these aggressive sales tactics and the pressures from higher up to open fraudulent accounts.[20] If the leaders knew about these warnings, they clearly didn't heed them.

Digital technology advances may also be a boon for established players. Although a single blog post can shatter confidence in a firm's product quality, the same social media also allows for broader and deeper direct connections with customers—as well as timely corrections when errors occur. Low-cost competitors could emerge from Malaysia, Pakistan, or Israel, but their appeal to customers also offers a clear market signal for a truly vigilant organization, giving it time to respond.

The vexing problem of an overwhelming number of cybersecurity alerts, for example, may be solved with artificial intelligence that filters out false alarms so that technicians can concentrate on genuine warnings. An Alphabet spin-out called Chronicle[21] is part of an immune system to help organizations defend against cyberattacks before they can infiltrate internal networks and cause damage. Cybersecurity remains a cat-and-mouse game, and computers will have to play the role of cat more aggressively. This challenge extends to other kinds of crime detection and prevention as well, such as international drug cartels, human trafficking, and terrorism.[22] Artificial intelligence models today disrupt criminal networks and help find gang leaders and their accomplices. Counterterrorism teams routinely scan social media communications and other sources using clever algorithms that find early warnings. Vigilant leaders must also learn to harness these new digital capabilities to protect their firms and business interests. The same kind of artificial intelligence (AI) techniques used in counterterrorism can be used by companies to sense and react faster. Facebook, for example, has an internal database to track rivals, including young start-ups that could become a threat or opportunity. This database gives Facebook an early-bird data stream of promising new service features it might acquire or copy.[23]

Navigating Digital Turbulence with Vigilance

Why are some firms more adept at anticipating the opportunities and threats from digital and market turbulence, while others struggle to keep up? This question is at the crux of this book. The answer we will develop is that the winners have superior vigilance capabilities that are exercised through deeply embedded organizational processes. Although nearly every organization will be blindsided sometimes, the vigilant firms are better prepared.

Successful incumbents will not allow themselves to be lulled into complacency. They know that an "it is not going to happen to us" narrative offers false comfort. They may tell themselves that weaker competitors are more vulnerable to digital disruption and that, as the tale goes, "I don't have to outrun the bear … I just have to outrun you." They also realize that this strategy only buys them time: if the bear keeps running, it will catch you too. To avoid complacency, strategic leaders keep three principles in mind:

1: Paying attention is a deliberate act. Vigilant organizations carefully manage which of the bewildering array of external and internal issues they need to attend to and which can be ignored. They know that attention is the scarcest of all organizational resources because it constrains the capacity to focus on and respond to pressing issues each day. To pay attention to everything is equivalent to paying attention to nothing. As Nobel Laureate Herbert Simon wrote in 1971, "A wealth of information creates a poverty of attention. More information is not always a good thing if it leads to blinkered thinking and analysis paralysis."[24]

So how should leaders allocate their own and their organizations' limited attention? Within vulnerable firms, leaders direct most of their attention toward current operations to meet short-term performance targets, using any scarce remaining time to react to unexpected events, unwelcome surprises, or internal political tensions of the moment. These leaders seldom have much time left to reflect on the bigger picture and discuss what really matters in the future. Hence their response to unexpected change tends to be weak, fragmented, and rushed. This is why canvassing the bigger picture ahead of time truly matters: it can foster organizational preparedness when internal or external challenges suddenly appear.

When Alan Mulally took over as the CEO of Ford in 2006, for example, the survival of the company was in question. It had lost 25 percent of its market share in the prior seven years and was hemorrhaging cash. One of Mulally's first moves was to bring a strong dose of candor and vigilance into the weekly Thursday morning meeting of the senior leadership team. Before Mulally took over, these weekly meetings were reputed to be like mortal combat zones.[25] Executives looked for points of vulnerability among their peers and emphasized their own self-preservation over collaboration. Each carefully vetted and rehearsed their presentations in advance so there would be no surprises.

Mulally changed all that. He started each meeting by inviting senior executives to share their internal problems, as well as any anomalies they were noticing in the external environment. What troubles are you facing? What puzzles you and why? What does it mean for us and what can we do? This frank approach so completely altered the tone of the meetings that at first no one volunteered to share any faint external stirrings that puzzled them. As Mulally persisted in surfacing obstacles and anomalies and shared his own concerns, he pushed his leadership team to widen their lenses and expand their viewpoints. In due time, they became brutally honest about their reality and more open to new ideas from outside.

Mark Fields, who succeeded Mulally as CEO, noted in a speech honoring his predecessor that talking about problems was viewed as a sign of weakness within the old Ford culture. Mulally challenged this macho view and argued it was actually a sign of strength to recognize problems early, including collective ignorance, so that leaders can tackle them honestly. This profound change in perspective paid dividends at Ford, which was facing a $17 billion loss in 2006 when Alan Mulally came on board. Three years later, the company approached Wall Street with a financing plan and bankers gave it $23.5 billion in new loans as a clear sign of confidence.[26]

2: To act faster, you need a new perspective on speed. Once organizations have sensed an incipient change and are starting to understand what it could mean, the question becomes, what action to take? In the maelstrom of digital turbulence, speed is an especially useful creed. First, delays usually increase the damages and limit the response range if rivals get there ahead of you. Second, seeing sooner allows more time to develop strategic options ready to be exercised when the time is right—thus avoiding hasty, irreversible investments. Finally, there are well-documented competitive benefits from moving first to establish preemptive positions or forestall negative snowball effects when scandals break.[27]

Just because the clock of business is whirring ever faster doesn't mean that leaders must operate in haste. Acting faster than rivals is about being ready for action when needed, and this starts with early detection and learning through probing questions and exploratory forays. But only after sufficient clarity has been achieved about key issues can leaders orchestrate better organizational preparedness in the form of multiple options and contingency plans. The aim of seeing sooner is to have more degrees of freedom later, when quick or bold actions are called for, without being boxed in by rivals' moves. Most managers prefer to act on their own terms rather than be forced to react to someone else's initiative.

3: Vigilance capabilities foster agility. Organizations at the bleeding edge of digital turbulence are moving from a comfortable and known risky environment (in which decision outcomes can be specified and probabilities calculated), toward the deep uncertainty of unknown unknowns.[28] Familiar and predictable environments usually can be navigated by "doing things right" and using ordinary capabilities for the proficient execution of current processes,[29] such as supply chain management, executing routine transactions, and delivering reliable performance. To navigate deep uncertainty, in contrast, requires a more vigilant toolkit based on three dynamic capabilities: *sensing* change sooner than rivals, *seizing* opportunities more effectively, and *transforming* the organization as needed to stay ahead. Vigilance results when companies master these three dynamic capabilities by operationalizing them through a host of sub-capabilities.[30]

With superior dynamic capabilities, an organization is more agile when turbulence is high. Agility means being able to move quickly and shift resources adroitly to higher-value activities sooner than rivals. For example, agile methods are activated when a scrum is formed to tackle an emerging opportunity or address a recent threat. A small team or scrum of three to nine people is assembled with all the diverse skills needed to carry out the initiative. These are self-managing teams, following a transparent process, using design-thinking methods to develop and test prototype solutions and learn quickly. They are the antithesis of cumbersome, top-down innovation processes with repetitive meetings, extensive documentation, and many impediments to progress. In the rest of the book, we dive deeper into the full meaning of vigilance and the benefits of seeing sooner and acting faster.[31]

Rewarding Vigilance

Do firms that invest in building vigilance capabilities subsequently outperform other firms? A recent study tackled this question and answered with a resounding yes.[32] Researchers assessed the vigilance capability (or what they called the *future preparedness*) of eighty-five European

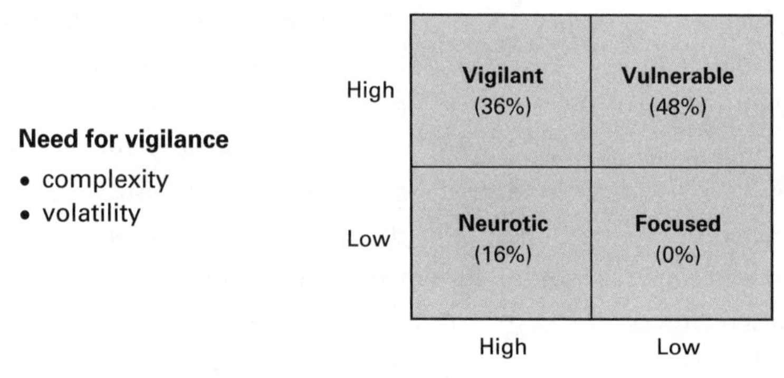

Figure 1.2
Does vigilance pay? How investments in foresight influence performance
Source: The bar chart reflects data published by Rohrbeck and Kum[33]

multinationals in 2008, then waited eight years to assess the benefits of vigilance. This time lag was long enough to see tangible performance differences, measured in terms of the economic success of these same firms in 2015. Building on our earlier research,[34] this research team measured an organization's foresight capability relative to its need for vigilance (see figure 1.2). Multiple metrics were used to score each firm along these two dimensions; reasonable cutoffs were then used to place firms in one the four categories shown. This classification yielded 36 percent of firms deemed vigilant by the researchers in 2008. This top group in turn was 33 percent more profitable in 2015 (measured as earnings before interest, tax, depreciation, and amortization) than the rest of the firms. The vigilant companies also had a 75 percent gain in their market capitalization since 2008, whereas vulnerable firms gained only 38 percent over the same seven-year period.

Vigilant firms outperform their rivals by successfully balancing stability with dynamism. The economic rewards are gained in many ways. Here are some of the diverse paths to profitability enabled by their vigilance:[35]

- *Capturing gains sooner:* Vanguard, a nonprofit investment giant, was an early adopter of artificial intelligence to bring financial guidance to its customers at a lower cost. Its Personal Advisor Services systems automates tasks such as rebalancing a portfolio toward a target mix or providing goals-based forecasts in real time, while human advisers take on higher-value activities and serve as "investing coaches." The payoff has been maintaining lower costs while keeping customer satisfaction high. In 2018, Vanguard had over $5 trillion in assets under management, serving over twenty million investors in 170 countries.[36]
- *Slowing down rivals:* In 2011, Kiva's robots and inventory-management systems were breakthroughs. Items to be shipped were brought to a packer near the truck door, rather than having a human picker/packer find the item in the warehouse. Amazon was early to see the potential of this system and purchased the company. This prevented the sale of Kiva's products to competitors, who now had to find other ways to keep up with the ever-increasing need for speed spurred by automation. These erstwhile competitors had to find other options; the only problem was that there were no other options.[37]
- *Gaining a first-mover advantage:* Print Yellow Pages has long been an endangered species as its economic model deteriorated compared with Internet search engines. But Telecom New Zealand anticipated this ahead of others and sold its directories business in 2007 for a handsome nine-times-revenue multiple. Most other telecom companies were slow to see this coming and act accordingly; they held on until their directories were nearly worthless.

- *Turning threats into opportunities:* The next chapter examines how Adobe Inc foresaw that its business model of selling boxed software would be imperiled and smartly moved to a cloud-based subscription service ahead of its rivals. We examine this turnaround in some detail in the next chapter to show what it takes for leaders to build a vigilant organization.

Future Shock Is Still Here

Few people are adept at seeing what's coming around the corner, and even fewer know how to act effectively on this information. Alvin Toffler was an exception. In 1970, he predicted that society's accelerating pace of change would initially be disruptive, then become the new normal and continue to intensify. Toffler labeled this as "future shock … a dizzying disorientation brought on by the premature arrival of the future" and characterized by "confusional breakdowns" throughout every structure of society.[38] Nearly fifty years later, it's no longer the premature arrival of but the belated responses to the future that sickens executive teams, their boards, investors, and society.

The leadership challenge of the future will be to blend big data, machine learning, human judgment, and artificial intelligence in ways that create a semi-sustainable competitive advantage—and do so in a way that is proactive, not reactive.[39] The future will belong to vigilant enterprises that combine soft and hard approaches to seeing sooner and acting faster. This blending of qualitative (soft) and quantitative (hard) approaches is difficult because of a cultural divide in our education system between the humanities and the sciences, which is reinforced through occupational specialization and acculturation. The reality is that vigilant organizations will need to have many arrows in their quivers, and vigilant leaders will employ the power of diverse approaches to seeing sooner and acting faster.

2 Vigilant Organizations

> If you can connect all the dots between what you see today and where you want to go, then it is probably not ambitious enough or aspirational enough.
>
> —Shantanu Narayen, CEO, Adobe Inc.[1]

By 2008, Adobe Inc.'s image-editing program Photoshop had gained the rare status of a product that was also a verb, like Xerox or Google.[2] Yet the new CEO, Shantanu Narayen, was uneasy: growth was slowing and revenue gains were mostly from price increases. Furthermore, a recession was brewing and the ubiquity of smartphones meant that everyone could be their own photo editor. As one executive noted, "growth in our business did not match what was happening all around us. Visual expression was on the rise everywhere ... there was more digital photography being shared, not less, and there was more video on websites, not less. Our business was solid, but it was not growing."[3]

The great recession of 2008 was a punishing time for Adobe Inc. By March 2009, its stock was down 60 percent in six months, and there were rumors that Microsoft might acquire it. During the next two years the leadership team embarked on two digital transformation initiatives: first, the 2009 acquisition of Omniture, to enable entry into the emerging field of digital marketing; and, second, a change to its Adobe Creative Suite business model, from boxed software on a disc—which gave the user a perpetual license to use the software—to a cloud-based subscription providing access for as little as fifty dollars per month. Announced in November 2011, the move to a software-as-a-service model was met with outrage from loyalist customers, who didn't like the idea of "renting" rather than owning and storing their creative content in the cloud.[4]

Because subscription revenue comes in over time, this change resulted in net income dropping by 65 percent in 2013. However, Adobe Inc.'s belief that lowering the price would attract new users—who wouldn't pay $700 or more for a software license—was proven resoundingly correct. By 2019, there were over 10 million Photoshop subscribers, Adobe Inc.'s stock price had tripled on expectations of continued growth, and its market capitalization exceeded $100 billion. The speed of the shift to subscriptions is shown in figure 2.1.

The move to a cloud-based subscription model was explained by Narayen in the spring of 2011: "We fundamentally believed we had to reimagine the creative process. If we reimagined the creative process as

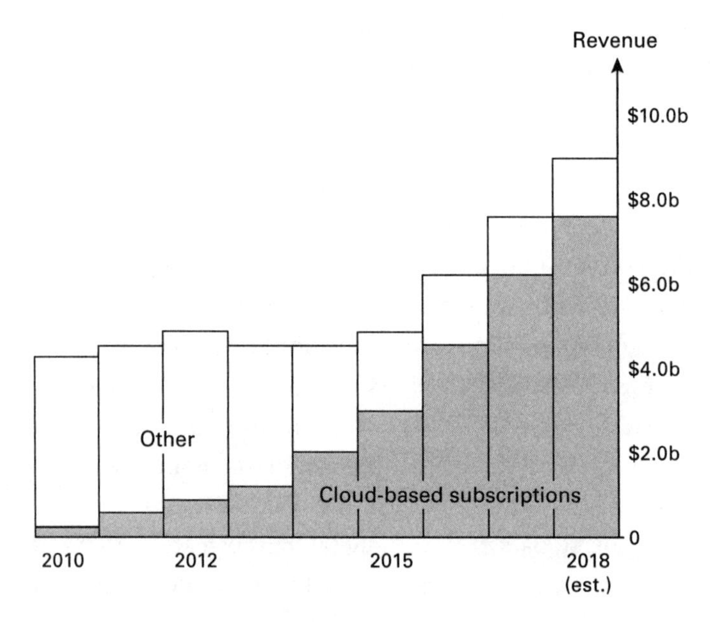

Figure 2.1
Adobe Inc.'s shift to the cloud
Source: news.adobe.com

a combination of the desktop, mobile and services—the manifestation was Creative Cloud. We also wanted to attract new customers to the platform. The next generation of customers would be far more familiar with the subscription model—they pay $40 per month for Dropbox. Determining 12 months in advance what customers might want and cramming as many features as we could into our products on a 12-month or 18-month cycle didn't work. We were shackling our engineers and allowing more nimble competitors to come in."[5]

Adobe Creative Cloud was launched in tandem with an upgrade of the Creative Suite in May 2012. In May 2013, the company advised its licensed software users that they would no longer provide software upgrades; further innovations would only be offered in the cloud. Cloud pricing was set to appeal to younger creatives struggling to pay the up-front price of a perpetual license. The needs of these creatives were evolving rapidly as they were becoming increasingly mobile, seeking connected workflows and a faster pace of innovation. Some Creative Cloud subscriptions for stand-alone creative products such as Photoshop were priced as low as twenty dollars per month.[6]

How did Adobe Inc. exercise vigilance and see the nascent opportunities in the cloud and digital marketing sooner—and act on them ahead of potential rivals such as Google, Salesforce, Oracle, and IBM? How was it able to overcome the forces pulling leadership teams to look inward and live in the moment? The key was to think holistically about the complete customer experience, rather than emphasize product features and functionality.

Like all vigilant organizations, Adobe Inc. was better than its rivals at dealing with the paralyzing effects of information overload and directed its scarce attention resources to the most important places. Such heightened vigilance requires a step change in thinking and behaving. Today's organizations need to emulate Formula One race car drivers, who have had to rewire their brains to drive differently.[7] The normal instinct of someone trying to drive *really fast* is to focus even more intently on the road just ahead. But the best Formula One drivers do just the opposite, looking intently out of the top third of their windshield. By focusing *farther ahead,* they see greater possibilities while being poised to avoid potential trouble.

Vigilance versus Vulnerability

Vigilant organizations like Adobe Inc. operate toward the vigilant end of the spectrum displayed in figure 2.2, and their leadership teams differ from those of more vulnerable rivals in the following ways:

- They exercise *vigilant leadership* with a deep sense of curiosity, embrace openness to diverse inputs, and project a willingness to play the long game.
- They *invest more in foresight* activities and adopt flexible, real-option approaches to help contain uncertainty.
- They adopt a *flexible and adaptive* process of *strategy making* that features outside-in and future-back approaches.

LOW *Vulnerable Organizations*	MEDIUM	HIGH *Vigilant Organizations*
• Short term focus • Conventional thinkers • Limited interest in outliers • Favor familiar settings	**Leadership posture**	• Play the long game • Willing to challenge assumptions • Seek diverse inputs • Involved in external networks
• Reactive posture • Formulaic/budget driven • Failures are errors	**Approach to foresight**	• Disciplined search • Flexible, real options approach • Experiment-to-learn
• Insideout thinking • Avoid uncertainty • Myopic and rigid process	**Strategy making**	• Outside-in thinking • Embrace uncertainty • Built-in flexibility/options
• Operationally focused • Information is silo bound • Weak signals left unattended	**Coordination and accountability**	• Focus on strategic accountability • Information shared across boundaries • Incentives for timely action

Figure 2.2
What is your vigilance quotient?

- There is *coordination and accountability* for acting on weak signals, reinforced by an organizational readiness to share information.

What do these attributes of vigilance look like in practice?

Leadership Posture

Shantanu Narayen, CEO of Adobe Inc., fits the profile of a vigilant leader. Exuding an eclectic curiosity, he engages with wide and deep networks outside his industry. A patient listener, he probes deeply and is comfortable with ambiguity, encouraging people to put their conflicting ideas on the table. This approach encourages new ideas and an openness to weak signals from within the organization.

Vigilance cannot be the sole purview of the CEO. It must pervade the senior leadership team to express the tone at the top and signal an openness to sensing and acting on early signals of threats and opportunities. The most influential members of this team are strong communicators who collaborate with other functions and serve as credible advisors to the CEO on all key decisions. Much more than advocates for the interests of their function or group, they can overcome the natural tendency toward isolated organizational silos that concentrate on immediate tasks.

Three qualities distinguish vigilant leaders from average executives: they focus externally and are open to diverse perspectives, they apply strategic foresight and probe for second-order effects, and they encourage others to explore widely by creating a culture of discovery.[8] The value of such a culture was summed up by Adobe Inc.'s CEO: "When you are trying to innovate and tackle a major transition it can feel impossible to connect all the dots from where you are to where you want to go, ... you have to create a culture that enables you to look at things with transparency, acknowledge failures, and course correct." To bring this culture to life, he argued, "we all have to focus on the storytelling aspects of what we are trying to accomplish ... creating narratives that people can engage with."[9]

Vigilant leadership teams create the kind of psychologically safe space required for employees at all levels to share perspectives about relevant issues outside their immediate domain. Employees deep down in the organization may be closer to weak signals at the edge of the organization and must believe they will receive an open hearing when raising concerns or suggesting ideas.

Vigilant leadership is more than visionary leadership. Although vision is an essential ingredient in any strategic transformation, there has to be a feasible path to realizing that vision. Otherwise, organizational acceptance will be an illusion and it will become ever harder to gain the alignment of people and processes that is essential to success.[10] Adobe Inc. confronted this challenge to its new strategy head on, as CFO Mark Garrett noted: "A lot of people didn't buy into the idea at the beginning. We knew that our revenues, earnings, and stock price were almost certain to drop during the transition. And we knew it was going to be a long, hard road. It really takes guts to make this sort of change—and that was what we had to come to terms with. Just how risky is it? Can we really pull it off? What happens if the stock goes down? What will customers, employees, investors, and board members think? We spent hours knee-deep in Excel spreadsheets modeling this out. This helped us get more comfortable with the idea of making the change—here's how much time it will take, here's what it will look like at the end."[11]

Approaches to Foresight

Before Adobe Inc. made its move to the cloud, it piloted a subscription-based pricing model for its Creative Suite software package in Australia, offering side-by-side, similar products under both a subscription model and the traditional pricing model. This experiment gave leadership confidence that the new business model could grow the base by attracting new users and increase the pace of upgrades by lowering the barriers to purchase. They also learned just how essential it would be to provide ongoing service value with continuous updates and new mobile apps.

Adobe Inc.'s experiment required careful planning, the investment of considerable time and money, and the commitment of leadership to follow the findings. They were systematically testing hypotheses about how to deliver a better customer experience. This exemplifies the approach to foresight we find in vigilant companies. Without the insights that arise from these investments, managers have to fall back on their experience or intuition, which may not be relevant.

Investments in foresight are not "one and done" quick fixes. Instead, they're built into a vigilance system. Here is a sampling of the foresight methods and tools used to strengthen vigilance that we will develop further in this book:

- The disciplined search for opportunities to innovate. A directed search will surface better-quality ideas than a scatter approach that sweeps up low-yield possibilities.
- A portfolio of continuous, small experiments that expands the repertoire of known patterns of response. This could include venturing units that integrate start-up initiatives.
- A centralized foresight unit or team, coordinating closely with the strategic planning group, marketing, and other forward-looking functions.
- Customer and supplier workshops or clinics aimed at understanding emerging customer preferences or possible vulnerabilities in the supply chain.
- Global scouting teams visiting innovation hubs such as Silicon Valley or new, growing market segments (such as young consumers) to gain fresh insight firsthand.
- The systematic cultivation of the extended network of partners, collaborators, suppliers, customers, and other external contacts. A large organization may have thousands of touchpoints in its total network that provide an early warning system.
- Strategic dashboards to monitor external scenarios and test the validity of core assumptions or track the progress of key projects.

These foresight investments produce direct, first-order consequences by providing deep insights for seeing sooner, testing hypotheses, and validating action options. Without these insights, a vulnerable organization has to rely on gut feeling and instinct and lacks an expansive repertoire of strategic responses to turbulence.

Strategy Making

Vigilant organizations regularly revisit their strategic priorities, choices, and direction and are better prepared to capitalize on rapid-fire strategic challenges. Their advantage comes from an outside-in approach that keeps them in tune with customers, partners, and competitors, and from a willingness to embrace uncertainty. This adaptive approach to strategy making was behind Adobe Inc.'s acquisition of Omniture and its decision to migrate to the cloud.

The impetus for the acquisition of Omniture came from a "blue sky" strategic retreat in 2008, during which Narayen and his leadership team tackled the slowing growth of their core business and the limited prospects for growth in their current portfolio of innovation initiatives. They knew they needed to push into market adjacencies with better future prospects. As Narayen observed, "Internally, there was a fundamental belief that where content was being created and managed was changing, where content was being consumed was changing, and where content was going to be monetized was changing. We also believed that data was going to become more important. Because we already had a huge presence in content creation, it was logical to broaden our lens to look at opportunities in each one of those areas."[12] Moving into the adjacent content-management space also meant entering the enterprise software market. Adobe Inc.'s DNA was encoded for desktop delivery, so many new capabilities were required to make this shift. Fortunately, its strong brand gave it permission to enter this adjacent market because it was already well regarded by the advertising agencies, publishers, and marketing departments using its packaged software. Some of the needed capabilities were obtained through the October 2009 acquisition of Omniture, the leading Internet marketing company and web analytics specialist. Omniture used a software-as-a-service approach to sell enterprise software to customers, such as web analysts and directors of marketing.[13]

The aim of Adobe Inc.'s digital marketing initiative was to become mission critical to digital marketers by transforming how they make, manage, measure, and optimize their marketing campaigns. The following quote from Mark Garrett, executive vice president and chief financial officer, illustrates how outside-in thinking was internalized within Adobe Inc. and its leadership team: "We wanted Adobe to be the key partner to the CMO, similar to the way Salesforce would say it's the key partner to the head of sales. And we were building out a complete solution. Salesforce, IBM, SAP, and Oracle were just realizing this was a big space, but no other company could say they were the key partners to the CMO. In Google's case, an enterprise customer like Nike might be leery about giving Google its digital marketing business and also buy key search words from it."[14]

The integration of Omniture accelerated the transfer of tacit knowledge about the latest advances in data analytics throughout Adobe Inc.[15] This kick-started the digital transformation of Adobe Systems and became a big selling point for current and prospective customers uncertain about how to compete in rapidly digitizing markets. To its credit, Adobe Inc. shared its best practices widely while learning from its customers about where and how to innovate its cloud offerings.

Another champion of an outside-in approach to finding and choosing strategies is Jeff Bezos, CEO of Amazon, who argues: "Rather than ask what we are good at, and what else we can do with that skill, you ask: Who are our customers? What do they need? And then you say you are going to give that to them regardless of whether we have the skills to do so."[16] He describes this as a "working backward" mentality to be instilled throughout the organization.

Successful strategic moves often begin with a leadership team collectively stepping outside its organization and seeing it through the eyes of customers, competitors, and other stakeholders, and then taking action.[17] This significant shift in stance requires a combination of humility and empathy that is often scarce, plus a willingness to scan widely and then entertain multiple perspectives. The necessary ingredient here is a deep sense of curiosity.

The strategy dialogue of vulnerable firms, in contrast, is more likely to start from the inside-out, with questions such as: What are we good at? What are our capabilities and technologies? And what else can we do with them? These are worthy questions, but they impose a narrow but comfortable frame and inadvertently constrain the subsequent dialogue.

Why is inside-out thinking pervasive in vulnerable firms? First, there is the seductive reinforcement of focusing on cutting costs and improving efficiency. If the performance benefits are quick to materialize, the quest for steady improvements in operations will be especially appealing. Second, there is the difficulty of balancing inherently competing priorities. Internal concerns about resource allocation, budgets, and turf wars are often most pressing. Lastly, the resource-based view of the firm implicitly endorses an inside-out approach to strategy making. This influential theory suggests that a firm's defensible competitive position comes mainly from its distinctive, hard-to-duplicate resources residing deep inside the organization. These resources bestow advantages that were cultivated slowly over time and may be hard for competitors to imitate. They need to be leveraged, and the job of management is to improve and exploit them for further advantage. Such inside-out thinking may work in stable times, but not amid turbulence.

The counterweight is the dynamic capabilities model mentioned earlier, the underpinnings of which reflect an evolutionary approach to *sensing* change, *seizing* opportunities, and *transforming* the organization. This model can be at odds with the "leverage your current capabilities" view, especially when the environment is changing rapidly or profoundly.

Coordination and Accountability

This fourth attribute of vigilant organizations enables the other three to flourish, setting the organizational conditions for success—or, when absent, setting up obstacles to sharing information and acting quickly. One compelling message from our research was that vulnerable firms suffer the ills of bureaucracy. Their hierarchical structures, along with the standardized rules and procedures needed to coordinate dispersed activities, come at a steep price:

- There is limited lateral sharing of information across functions and geographies, so it's hard to get a full picture of an ambiguous trend or an anomalous event.

- Bureaucracies are slow moving because many people have to be involved in decision-making.
- There are few incentives for taking timely action on ambiguous signals. Bureaucrats are instinctively risk-averse, and rules are designed to avoid mistakes.
- There is very little accountability for taking action on weak signals. Responsibility is diffused or absent and coordination suffers.

How then do vigilant organizations achieve the coordination and accountability needed to act faster? Whereas vulnerable organizations are built on the principle of decomposition of activities and functions, the agile, vigilant organization uses some variant of an *adhocracy* design, in which activities are combined and actions matter more than a person's position. In an adhocracy, probing the periphery for ideas, trying things out, and continually experimenting are more important than following the rules and avoiding responsibility.

The organizational processes used by vigilant firms are guided by design principles that encourage agility. First, there is someone in the leadership team with responsibility for collecting the paranoia, assembling the weak signals, and then interpreting their significance. These people may be located in centers of excellence, responsible for a broad domain of interest. Second, vigilance is a team sport, so groups may be formed into geographic units or participate in a core process, such as supply chain management or one that is given the mandate to explore the relevance of blockchain technologies. Third, there are procedures to identify and share compelling insights through staff communications and briefings or a common information system. The message is that collaboration and communication, as opposed to competition and isolation, is a better way to see sooner and act faster.

Accountability for sensing weak signals early can be fostered by establishing clear metrics that focus the organization on specific scanning tasks and then tying group or individual incentives to successfully executing the necessary activities. Narayen was a strong believer in planting flags in the ground and setting three-year targets. This sends a strong internal signal about what matters and has influential signaling value for external stakeholders. It's fitting to conclude with a quote from the CFO of Adobe Inc., Mark Garrett, about how the company got the stock market to appreciate its strategy:

> We had prepared Wall Street for a significant drop in revenue and earnings in 2012. We also shared new metrics to help analysts measure the health of the business as we went through this transition. We shifted their focus toward the building blocks of the Creative Cloud business—subscriptions, annualized recurring revenues (ARR), average revenue per user, and revenue that was contracted and either deferred or in backlog (off-balance sheet). We gave them "markers"—for instance, we said we were going to reach 4 million subscribers in 2015 and build up ARR. As the switch-over progressed, toward the end of 2013, investors became intrigued and started asking about longer-term objectives. So we projected the compound annual growth rate and earnings per share out three years and shared those metrics.[18]

Vulnerable or Vigilant?

Most firms are closer to the vulnerable end of the spectrum in figure 2.2. Their operational focus favors the here and now of responding to immediate competitive moves and pressing customer demands while cutting costs and pursuing efficiency gains. This posture is increasingly perilous when there is digital turbulence and mounting uncertainty.

Our clients want to know how they should focus their energy to become more vigilant. What specific actions will deliver the best results? We'll provide a detailed action agenda in chapter 7, based on what we've learned about how the four vigilance attributes work together to distinguish vigilant from vulnerable organizations.

The diagnostic tool in appendix A is designed to calibrate organizations on each of the four attributes. We've used this survey many times to help our clients find where they are on the vigilance spectrum. To provide context for their judgments—and answer the question of how big a gap there was between their score and that of superior performers—the survey was given to a sample of 118 firms.

The survey uses multiple questions to assess each of the four attributes of vigilance. We scored all the firms in the sample on these attributes and then estimated which attributes best explained their past performance in seeing sooner and acting faster (details of the analysis are described in appendix B). We found that *vigilant leadership* and *investments in foresight* were most highly correlated with performance. These two attributes are the drivers of vigilance and are the necessary conditions for success. The vigilance improvement program must begin here.

How an organization approaches strategy making and ensures coordination and accountability are enablers that leadership can use to make improvements. On their own they cannot overcome myopic leadership or inadequate investments in foresight. For example, an outside-in approach to strategy making is crucial, but what matters is how leadership implements the approach.

The challenge for leaders is to balance outside-in and inside-out thinking, independent of whether the issues happen to be internal ones or come from outside the organization. Outside-in thinking means moving away from the perspective that gave rise to the problem perception in the first place, which is often a parochial or self-centered view. Especially in digitally turbulent environments, analysis must begin by reframing the issue from the outside in and then iterate back and forth between current concerns (see figure 2.3). Legendary Intel leader Andy Grove was a strong proponent of expansive outside-in thinking because it naturally challenges the initial problem framing. Constantly on the lookout for strategic inflection points in the marketplace, he prodded his leadership team to anticipate and respond to game-changing shifts in the computer chip industry. After being appointed CEO, he provocatively asked his management team: "Suppose that we had just acquired Intel: would we stay on the same path?"[19] This profound outside-in question eventually led the management team away from memory storage and toward semiconductors.

Vigilant leaders distinguish between situations that can be reliably predicted—such as demographic trends, business cycles, regulatory changes, or technological trends—and external uncertainties that neither their organizations nor outside experts fully comprehend. The latter would include the financial crisis that erupted in 2008, the political

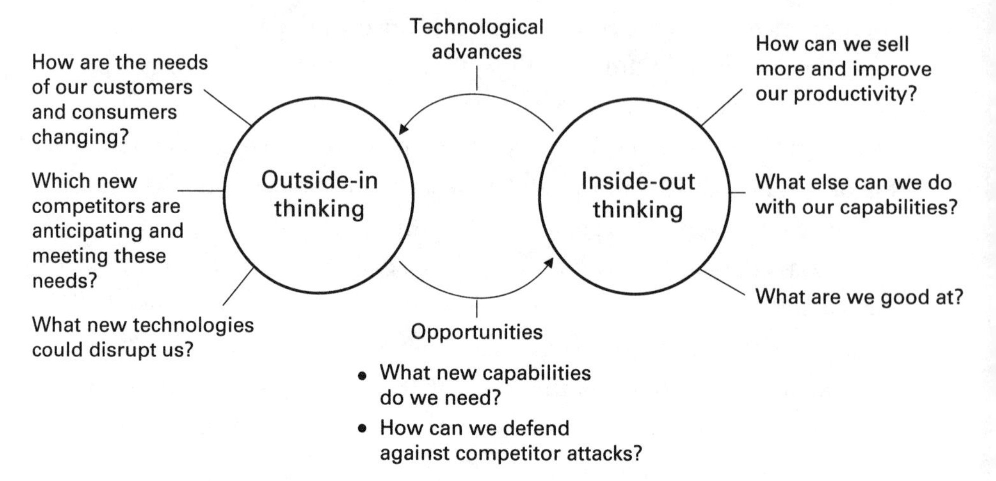

Figure 2.3
Integrating outside-in and inside-out thinking

upheavals in the Middle East, the development of cloud computing, and the promise of machine learning. In an uncertain and ambiguous world, the focus should be on asking better questions, seeking new insights, accelerating learning, and creating responsive organizations. Strengthening and employing agile organizational capabilities is much more important than simply emulating best practices.

Assessing Your Vigilance Quotient

During the past decade, we have worked with scores of global firms to help their management teams become more vigilant. We usually start the engagement by asking the leadership team members to recall significant instances in the past five years in which they were too late in seeing key trends or major turning points. Then we also ask for examples of times their organization was ahead of external changes or internal developments to complete the cells of figure 2.4. The goal is not finger-pointing or scapegoating, but to get a balanced picture of hits and misses to see if there are patterns in the team's blind spots or successes. Deeply understanding each episode may foreshadow sources of future problems and opportunities, as well new approaches.

	Threat	Opportunity
Seen in time to act	**Problem avoided**	Upside created
Seen too late and had to react	Scramble to catch up	Missed chances

Figure 2.4
Four categories of hits and misses

One medical diagnostics company specializing in high-end microscopes was very early to market with a high-definition microscope—but very late to see that cheaper, low-end in vitro diagnostics were eating into its market share. Another company, Olympus, was a highly regarded player in the camera business and was well ahead of the market in high-quality underwater cameras. These devices require a very robust design so that they can withstand intense pressure while also preventing salt erosion and water inundation. This capable company in terms of external sensing later suffered from a major internal accounting scandal that was seen too late and that badly damaged its reputation and financial performance.

How could the same company see one complex external opportunity quickly but be blind to a long-festering financial fraud within its own ranks? How could it see one type of external opportunity and not another? The fact is that vigilance is not equally distributed throughout an organization. Such inconsistencies arise because no organization is monolithic; some parts may be good at operations, others at innovation, and yet others at seeing around corners. As a result, the overall organization may not be the best unit of analysis, especially if it's large and multidivisional. Vigilance should be measured granularly because it varies greatly across teams, functions, departments, and business units.

Another electronics company we worked with seldom missed anything related to its technology but was usually late seeing shifts in consumer preferences regarding packaging, colors, weight, ease of use, and so on. These consumer issues actually had more impact on sales and net margins than the quality of the technology, but the engineering-centric culture of the firm weakened the deep understanding of "softer" consumer issues. They often left these *secondary issues*, as they termed them, to partners who often made higher margins on accessories or complements than the company did on its core product. To consumers, these additional features mattered a great deal and often were primary factors in their purchasing decisions.

Once missed threats and opportunities have been honestly surfaced from the past, the next step is for leaders to dig into the reasons why they missed them while seeing others. This can be a sensitive exercise politically and needs careful handling. For example, a European maker of reusable endoscopes—a common medical device used to examine the internal organs of the body—did not sufficiently recognize the threat of disposable alternatives until it was too late to defend its strong position. Reusable endoscopes are hard to clean properly, making disposability a major benefit for surgical teams. One reason this firm did not pay much attention to users' frustrations is that it viewed itself as a manufacturer of *reusable* endoscopes. This narrow mental frame and self-definition created myopia and caused the organization to miss multiple emerging technologies that in combination made disposable endoscopes viable substitutes. The company also failed to monitor improvements in the production process that allowed rivals to bring less-expensive devices to market faster.

After surfacing a representative cross section of hits and misses, we then work with leadership teams to uncover any recurring patterns and identify deep causes of why the organization may still be vulnerable. We look for examples of early as well as late detection. One of us worked at Royal Dutch/Shell planning group in London when it became well-known for its stellar capability in scenario planning. This approach helped Shell understand megatrends and uncertainties better than most of its rivals. Still, when it came to understanding consumer sentiments and addressing them through effective use of media, Shell was less than stellar. Several megafailures in the late 1990s and early 2000s revealed a pattern of significant weakness in its societal understanding of media, special-interest groups, and public relations.

In one instance, for example, Greenpeace strongly protested Shell's plan to sink its massive defunct Brent Spar platform in the North Sea in 1995. The social uproar orchestrated by Greenpeace resulted in consumer boycotts in Germany and the Benelux countries, forcing Shell to drop its plan. In another case, Shell funded schools and other socially worthy projects in Nigeria, but the company didn't monitor the fund flows closely enough; when too much landed in the pockets of corrupt politicians in Lagos, Shell was embroiled in embarrassing lawsuits. In a third case, Shell overstated its underground oil and gas reserves by using overly favorable estimation techniques at a time when public distrust of companies was high. Its aggressive reserve estimates happened just on the heels of corporate scandals involving Tyco, WorldCom, Enron, and others. The inflated underground reserve estimates cost Shell's very powerful head of exploration and production his job. Clearly, there was a serious gap in Shell's vigilance when it came to perceptions about outsiders, including the eagerness of various external parties to hold the company to account.

We would be remiss if we didn't mention several challenges that can come from leaders performing inside audits of "vigilance failures." If these failure reviews are not done openly and managed sensitively, they may become scapegoating exercises that produce the opposite of vigilance due to underlings distorting or hiding information. Also, *hindsight bias* can easily mislead and produce superficial advice aimed at avoiding the same mistake again, rather than tackling the deeper organizational issues. Such surface patches can create a false confidence and make it more likely that the organization will miss other important signals. Organizations can avoid these risks by keeping sight of the goal, which is to identify deeper patterns and find broader areas for improvement—not to settle old scores or close the barn door well after the horses have escaped.

Done properly, assessing past vigilance successes and failures should energize a leadership team. To extract broadly applicable lessons, the team must try to understand and agree on cases in which important signals were missed in the past, as well as times it was ahead of the curve. This will help leaders focus the organization on what matters most so that team members don't feel helpless or confused by the dizzying volume of noise coming at them every day.

The next chapter explores the collective attention of the leadership team, which may be the scarcest resource of all. Just as each person has a limited attention range and time reservoir, so does an organization. The role of leaders is to assess where the collective attention may be deficient and then focus the organization's mindset accordingly. Missed signals may reflect excessive filtering and a misallocation of attention, just as seeing opportunities sooner suggests a proper focus and scanning approach. Organizational attention should be managed strategically at the individual, group, and broader organization levels.

3 Managing Organizational Attention

A wealth of information creates a poverty of attention.

—Herbert Simon[1]

In an increasingly complex world, the scarcest collective resource of the modern leadership team may be the most important: attention. In vigilant companies, leadership attention is leveraged for greater agility and advantage, whereas in vulnerable companies, misdirected attention creates blind spots, myopia, and delayed reactions.

This explains why the Polaroid Corporation was so slow to move from analog to digital imaging.[2] The company's leadership was so intently focused on the instant film side of the business that they initially missed the shift to filmless digital cameras. Their collective mental template was dominated by a commitment to a "razor/razor blades" business model, whereby profits for the camera company were to be made by selling quickly developed film cartridges. The camera in this case was like a razor—simply a means to an end. When digital cameras became a viable and popular alternative, Polaroid struggled to catch up. In the end, its struggles were in vain, and the company filed for bankruptcy twice between 2001 and 2009.

A turbulent business environment is forcing many traditional companies to challenge their business models. Almost every big oil and gas company's leadership, for example, is refocusing its attention toward opportunities in clean energy while also harvesting their still-profitable hydrocarbon businesses. At present, a second wave of clean energy strategies is afoot that may dwarf the previous one. Fueled by the growth of electric cars and the changing economics of renewables, leading players are shifting their view of the future. Shell, for example, bought a Dutch car-charging network; Norway's Statoil (now called Equinor) deployed the world's first floating windfarm in 2017; and France's Total took a 23 percent stake in Eren, a renewable energy firm.

Across many other industries, senior leadership teams face two pressing issues: (1) how to expand their collective attention resource so that it doesn't become a constraint on the ability to see sooner and (2) how to best allocate such expanded attention to spotting weak signals of threats and opportunities from both inside and outside the firm's boundaries.

The contrast between *vigilant* and *vulnerable* organizations in tackling these challenges is striking. The leadership of vulnerable organizations—such as Toys"R"Us and so many others that have filed for bankruptcy—remains preoccupied with operational concerns and coping with daily pressures. As a result, most of their time is spent reacting to events, rather than shaping them and anticipating avoidable crises. This treadmill of firefighting can easily become a downward spiral, with less and less time available to sense weak signals from the periphery or to probe them more deeply. In contrast, vigilant organizations recognize full well the limits of managerial attention and stay one step ahead.

The Psychological Limits of Individual Attention

Every leadership team is a collection of individuals, and as human beings, each member of the team has an inherently limited ability to pay attention, to absorb and process information through their mental filters. Fortunately, a good deal is known about these limits and how they can be overcome.

Seeing around corners is no easy task. We *all* have limited mental resources and implicitly block signals that are not deemed relevant. The problem is that in isolation, some signals are so disjointed or seemingly meaningless that it's hard to recognize the underlying threads that connect them. The challenge here is not just cognitive, but also emotional; there may be things we do not want to know or see, leading to "willful blindness."[3]

Such blindness tragically struck Joline Gutierrez Krueger, a seasoned journalist with New Mexico's largest newspaper. This state has one of the highest opioid overdose rates in the United States, and Krueger had extensively covered the opioid crisis after it began to strike returning Vietnam veterans in the late 1970s.[4] Somehow, however, she completely missed both weak and strong signals of trouble in her very own home. Her oldest son landed in the local hospital due to an opioid overdose and was barely resuscitated after his heart stopped; he died shortly thereafter. As a trained reporter, she asked herself how her "mom radar" could have been so far off that she didn't sense the problem. She blamed herself for having been the "educational mom" who lectured her kids about the dangers of drugs but never noticed what was happening in her own household. This tragic episode, with its multiple causal factors, is emblematic of the challenges we all face in allocating our limited attention to what really matters.

Much of the research on attention has emphasized individual constraints on how much (or little) information human beings can process at any one time,[5] the varying degrees of control we have over what we notice, and the importance of motivation or *directed interest*. We'll discuss four valuable insights from this body of work that can help improve a leadership team's processing power.

1. *Attention is a filtering mechanism for balancing our internal capacity with external demands.* This capacity is not completely fixed: mental effort can rise if we are interested, curious, or need to concentrate intently. Mental filtering starts by processing weak signals and all the other stimuli that bombard us in every waking moment. Once stimuli present themselves from outside—or from within, through reasoning or imagination—the question becomes how to prioritize them. Clearly, our personal interests and dispositions influence what we pay attention to. If an urgent phone call is taken at work about a

family issue at home, that quickly becomes our priority. It also matters how much emotion we have invested in the issue at hand and what else is competing for our attention.

Managerial work is often fragmentary and fast-paced, usually initiated by outside forces, putting executives in highly reactive modes.[6] Just as the study of history has been summarized as "one damn thing after another," the same can be said of managerial work. The array of problems vying for attention can be dizzying, and CEOs vary considerably in how and where they spend their time.[7] At some point, we run out of mental capacity to handle it all, triggering a reevaluation of what is more or less important. Some attention-allocation heuristics may be preprogrammed, in the form of corporate policies, priorities set by the boss, and your own to-do lists, but many are determined in the moment. The end result of filtering is the generation of a set of possible activities, followed by concrete responses or actions.

2. *What we see depends on what we expect to see.* In human vision, the *periphery* is that fuzzy zone outside the area we're paying close attention to. As a result, for individuals as well as organizations, any weak signal emerging from the periphery is usually difficult to see, hard to comprehend, and ambiguous in terms of prescribed actions. Peripheral vision involves an interplay among sensing, interpreting, and probing, and what we see is strongly influenced by *what we expect to see.*[8] Individuals are often so intent on the task at hand that they become oblivious to a significant change in the environment because it's outside their focus. This was brilliantly depicted in the famous "gorilla walking through a basketball game" video experiment.[9] The viewer is asked to count how often a team wearing white shirts passes the ball among its players; at the same time, a team wearing red shirts is likewise passing a basketball on the court, also just among its players. Midway through, a man dressed in a gorilla outfit walks through the scene without disturbing any of the players. Although he even stops briefly to beat his chest, about half of viewers fail to notice him. They are just too busy counting passes.

3. The *ability to sense weak signals can be enhanced.* There is usually a cost involved in strengthening our peripheral vision, and companies must commit resources and make the development of these capabilities a priority. The challenge for both individuals and organizations is to find the right balance between focal and peripheral vision. Consider Bill Bradley's college basketball career before he became a US senator: Bradley possessed an uncanny sense of awareness of other players around him on the court. His peripheral vision was found to be outside an ophthalmologist's 180-degree scale, with a range of 195 degrees horizontally. Although he may have been naturally gifted, he also cultivated his peripheral vision as a child. When walking down the sidewalk, he would look ahead, keep his head straight and then try to identify items in the shop windows to the right or left. Later, he would stand at various places on the basketball court with his back to the basket, turn quickly, and try to shoot blind as the net appeared in his visual periphery. In the end, he honed what one reporter described as a superb "sense of where you are."[10]
4. *Attention can be (re)directed.* Because we can focus intently on only a few points of vulnerability or possibility at a time, weak signals of impending threat or opportunity need to be separated from the surrounding distracting noise. This entails asking probing and guiding questions about a few areas of high priority. Shell's CEO did this by asking his team this question: "Pushed to the extreme, how quickly could electric vehicles come?"[11] His attention had been caught by an alarming anomaly between 2014 and 2016, during which oil prices fell while electric vehicles *doubled* in global sales from 323,000 to 753,000 units per year. In the six years preceding 2016, the price of lithium-ion batteries used in electric cars had dropped 73 percent. Shell's head of planning characterized the resulting challenges facing the company as "radical uncertainty."[12]

   To help leadership grasp the situation fully, his group laid out the range of plausible alternative futures by integrating two pivotal uncertainties to create four possible scenarios: (1) the total global demand for energy of any kind and (2) the likely penetration of alternative energy sources such as solar, wind, tidal waves, biomass and others that would reduce the demand for fossil fuels. A combination of low energy demand and high technological substitution describes a world in which demand for oil will peak around the mid-2020s. This scenario would cause the most upheaval for big oil companies, and Shell optimistically labeled it the *brave new world*. Shell actually has no idea which of the four combinations implied by its two guiding questions best describes the future, nor how quickly each scenario may arise. But its attention has been proactively broadened toward understanding the implications of digital and alternative energy technologies, preparing it to act as the future unfolds.

Expanding Leadership Attention

How can a leadership team expand its collective ability to pay attention to weak signals when each of its members is already overloaded—and digital turbulence is only increasing the load? We can think of the attention resource as being analogous to a sponge, with the amount of information that can be absorbed and acted upon constrained by the organization's *absorptive capacity*—the limit of its ability to recognize the significance of new information and assimilate it.[13] The view that there are inherent limits to what an organization can absorb has significant implications for managing attention. [14]

An organization's absorptive capacity depends on individual managers' capacities, prior knowledge, and ability to transfer their knowledge via "gatekeepers." *Someone* has to be the point person on an issue or topic, and these people are most effective when they know where to find complementary knowledge within their organization. To foster these connections, some vigilant firms have invested in "mini-Google" internal search engines to help people quickly find who knows what about a topic of interest.

Increasing the diversity of your team also expands its absorptive capacity. We all carry around in our heads a set of subconscious biases, assumptions, and beliefs about how the world works. These mental models help us make sense of the world, but when change is rapid, they often stand in the way of fully grasping what is happening. This problem is exacerbated when everyone in an organization has a similar mindset and responds in the same way. The result: corporate learning processes often reinforce shared perspectives while discounting or excluding others. Thus, the first step for increasing adaptability is to diversify the talent pool to include people who are not yet wedded to the status quo. Outsiders or closely connected partners, such as advertising agencies or consultants, can bring different life experiences and an openness to divergent information.

There is always the risk that a deeply rooted corporate DNA will reject "outside invaders" by launching internal "antibodies" in the form of defenders of the status quo and thought police. Vigilant leaders preserve the value of diverse insights by protecting outliers from being marginalized.[15] Vigilant organizations are more successful at working around the limits imposed by their absorptive capacity by doing one of two things: reducing the need to process information or increasing the total capacity to process information.[16] Let's examine each in turn.

Reducing the need. When an organization is in a stable state, there are well established routines for executing the strategy and getting work done. When exceptions or departures from plans occur, the problem is sent up the chain of command, and eventually a decision comes down. When advantages are fleeting due to an environment that is changing constantly, the rising turbulence will cause more exceptions to established procedures of a traditional command-and-control structure. One way to reduce overload from these exceptions is to add slack by extending lead times or adding inventory buffers and capacity. Slack usually comes at a cost, but it produces benefits as well, such as reducing the incessant demands on a leadership team to resolve pressing issues. Giving senior leaders more time to reflect at a distance may well be worth the seeming suboptimality of slack.

Another way to tackle the need to process information vertically, to avoid overloading the hierarchy, is to reorganize the enterprise into smaller, self-governing units. Apart from traditional decentralization in terms of regions or product groups, one part of an organization could be mostly responsible for current operations, with another group tasked with preparing it for the future. Such divide-and-conquer approaches will eventually require an integration mechanism as well, which is when ambidextrous leadership abilities become important. We'll look into these various possibilities further in chapter 7.

Increasing capacity. This is an alternative approach to balancing an organization's need to process information vertically with its limited attention and resources. Vertical capacity can be increased by investing in better information-processing systems, such as formalized reporting procedures. Overall capacity can also be increased by allowing more lateral communication across organizational boundaries, both inside the firm by using matrix structures and by improving partner relationships. This is one rationale for maintaining an internal R&D capability rather than relying completely on outside resources. An internal research group can orchestrate the technologies needed to support the company's growth strategy, but only if it has the knowledge needed to properly assess technological advances. This will require a deep external network to obtain early warnings, as well as the ability to manage collaborations and build relational capital.[17]

Valve Corporation is one company that has pushed both approaches to the limit to cope with a fast-changing digital environment.[18] Founded in 1996 by former Microsoft employees, Valve started by the entering video games business and then evolved into new markets through radical empowerment and team wisdom. Breaking with tradition, it hired new talent solely based on a proven ability to identify and capture new market opportunities. Employees move freely between departments to work on whatever interests them. Importantly, they do need to take ownership of their products and their mistakes.[19] Valve employees steer the company, armed with the power to green-light projects. People are not hired to fill specific job descriptions, but rather to constantly seek out the most valuable work that they can perform. The aim is to move away from functional silos, a top-down hierarchy, and routinized tasks in search of greater creativity, innovation, and bottom-up leadership.

Valve's basic assumption is that when it comes to sensing new market opportunities, *no one* has all the answers. Its employees have full autonomy to propose projects, recruit project teams, set budgets, set timelines, and ship products to customers. This polyarchic approach to flattening an organization is the opposite of a traditional autocracy, in which a chosen few powerful people decide everything. This bold new model has already yielded several important products, including Steam, which serves as a product platform for digital distribution rights management, broadcasting, and social networking. The initiative started out as the brainchild of a few employees, without any top-down planning, and the platform already contains over 400 million pieces of content created by users, generating significant additional licensing and transaction fees for the firm.

The unbounded creativity unleashed by such a polyarchy must still be coordinated to yield sensible strategic moves while avoiding duplication of effort and other cost inefficiencies. Valve achieves this through a countervailing force called *social proof*, whereby the creative ideas proposed by individuals must pass through several filters to move forward, including *the rule of three*. Following the adage that "first you must let chaos reign and then you must rein it in," a proposal must obtain the support of *at least three individuals* to be given a green light. At Valve, this would mean approval to start work and accessing a preset amount of funding without formal approval. However, because other staff are aware of the project, an informal method of social validation takes form, but without formal top-down oversight. In fast-changing environments, project management generally tends to be more flexible and bottom up—for example, by using scrum teams.[20] Creative or uncertain tasks—in such areas as software, R&D, or environmental scanning—need agile approaches in which the focus is on people (rather than formal procedures) to encourage informality, speed, and collaborative interactions with diverse contributors who may have relevant tacit knowledge.[21]

This informal, spontaneous approach creates greater motivation, broader transparency, and deeper internal dialog. Creativity tends to flourish when people are free to explore and propose ideas without the need for group approval, which often stifles mavericks or self-starters. As with effective brainstorming, in the beginning there should be very little evaluation of new ideas—just further elaboration and extension. Once enough good ideas have been surfaced, the group can embark on a full vetting of the pros and cons of each proposal while proactively avoiding the traps of groupthink, polarization, and hidden agendas. Because everyone knows what's happening broadly, with chances to discuss the project as it proceeds, ulterior motives can be sniffed out quickly. Valve has thus managed to strike a better balance between the strong external market focus of its internal entrepreneurs and the wisdom and control of the crowd at later stages.

Allocating Attention

Effectively distributing the scarce attention of a leadership team is a trial and error learning process. Our diagnostic survey (in appendix A) found that firms are 50 percent more likely to be surprised by significant events from outside the firm than from inside, such as fraud, discrimination, bribery, or reckless behavior. Large firms with a global reach report they are surprised by outside events more than twice per year. These are only measures of the frequency of surprises—not the magnitude of the impact of the surprise. Deconstructing the frequency and type of past surprises is a good place to start a productive conversation about the allocation of a team's collective attention to each of the four cells of the myopia matrix (see figure 3.1). It helps to include examples other than just digital misfires because the aim is to understand where the overall organizational system itself may be deficient. Each cause identified, and remedied thereafter, serves to strengthen the organization's capacity to be alert and stay ahead of trouble.

Each cell in the matrix is prone to attentional misfires that might have been avoided if closer attention had been paid. The key is to examine a representative sample of cases and treat them as informative historical stress tests. The challenge is to look inside those cases that will best

	Threats	Opportunities
External	*Short-sightedness*	*Tunnel vision*
Internal	*Willful blindness*	*Missed chances*

Figure 3.1
The myopia matrix

surface deeper systemic weakness. The following are some nontech examples for each cell:

- *Short-sightedness* partly explains why Mattel's Barbie doll lost one-third of her market to the edgy and hipper Bratz doll. One reason was that the driving force at the top was a financial one. Also, Mattel's organizational structure impeded the sharing of information among market segment teams. For example, Ken (the male doll) and Barbie were hardly on speaking terms, and Mattel saw too late that both were losing interest among older children.
- *Tunnel vision* is a common consequence of a narrow definition of the market served and an overemphasis on current operations. Many global food companies like Kraft Heinz[22] have suffered from this condition and were too slow to respond to the shift away from packaged foods toward the perimeter of the grocery store (where fresh vegetables, fruit, and meat are sold). They also seem to have missed the shift to meal kits such as those from Blue Apron, which are both convenient and can lower grocery spending. These symptoms suggest some deeper systemic vulnerabilities.
- *Willful blindness* happens when we become aware of something that we would rather not know and therefore ignore or unconsciously suppress. Widespread ignorance of internal threats generally arises not because they are secret or invisible, but because leaders turn a blind eye to them. When former Secretary of State George Schulz

joined the board of Theranos, he got his grandson a job there. But this seasoned leader then refused to believe his own grandson when he confided that Theranos's micro-blood-testing technology and claims were mostly lies. Schultz had played a key role in promoting the company, raising hundreds of millions for Theranos, and met weekly with CEO Elizabeth Holmes during 2014. His grandson's warnings were apparently so devastating for him that he fell victim to willful blindness and never looked into whether the blood-testing technology was real or fake.[23] Eventually Theranos imploded and went from being worth billions to going out of business, with court cases piling up and many careers shattered.[24]

- *Missed chances* can occur when the organization's internal attention is focused narrowly without sufficient slack to explore opportunities at the periphery. Walmart prospered for years by placing cost-cutting at the center of its strategy. Yet only when the firm made environmental sustainability a major priority did it realize that being environmental-minded could actually reduce costs significantly. Why were these opportunities missed at first? Although seeing such opportunities late is better than never, Walmart's intense focus on saving costs made them slow to recognize the counterintuitive finding that being socially conscious makes for good business in terms of efficiency as well.[25]

The inherent simplicity of our myopia matrix makes it a useful departure point for a strategic dialogue. The categories are familiar and easy to grasp at an intuitive level. This is especially true for the threat versus opportunity distinction. In practice, both are more like tendencies than absolutes because this dimension is a continuum. Threats have a negative connotation, with the expectation of loss without gain and the likelihood of trouble if ignored. Judging a past surprise or a future issue as a threat leads to feelings of loss of control because others are imposing constraints. Conversely, opportunities have positive resonance because they imply a greater upside potential. The opportunity label highlights the upside (in terms of probabilities and payoffs), with the risk that downside risks are overshadowed or ignored.

Judging whether a past surprise was internal or external in origin is also dependent on people's perceptions and frames. Some cases will be clearly external in nature—such as a new piece of legislation being introduced (without your company having had anything to do with it). Likewise, an emerging technology may gain ground in a setting far removed from your own company, such as the Internet in its early days or ongoing innovations by Google, Amazon, or other master disrupters. Conversely, there will be cases in which the origin of an issue of concern is clearly internal, such as a fraudulent action by an employee or a manager treating employees inappropriately. Many cases will be more mixed. Cyberhacking, for example, is a pervasive external threat, but if your firewalls are breached, then it may be deemed an internal failure by the media or the board.

Even when the triggering events are clearly external, the way the issue plays out over time introduces ambiguity. Suppose an external supplier signals to one of your employees that it would welcome more business from your firm and will make it worth their while. The possible wrongdoing here—namely, bribery and kickbacks—is clearly external, but its occurrence will at first only be known to one internal employee. If this person takes the bait and acts illegally, the problem will be viewed as the actions of a rogue employee, and if other employees get drawn in without anyone blowing the whistle, the issue may be viewed as internal in origin by reporters or the law, ignoring the reality that the precipitating event was external to the organization.

Avoiding past unpleasant surprises is only a starting point for finding a better distribution of collective attention because it is susceptible to two problems. First and foremost is the danger that recent traumas will dominate the span of attention at the expense of other challenges. For example, in the wake of its bogus accounts scandal, it's understandable that Wells Fargo Bank focused first on repairing its toxic sales culture and then shifted to rebuilding trust.[26] But it's likely that these changes underway will only make it look like every other bank and possibly endanger the bank's ability to return to its previous growth path. It's never easy to turn a heavy blow into an opportunity, but almost any crisis also opens the door to dramatic changes in culture, business models, and competitive moves. This should not be ignored by Wells Fargo or by others encountering scandal or malfeasance.

The second challenge is that people cope with future threats by engaging in wishful thinking or resigning their futures to fate, both of which restrict the amount of information they explore and solutions they consider. In contrast, framing an issue as an opportunity results in a more open information search and in more explicit evaluations of options. How issues are labeled and how they are described greatly influences people's attention.[27] Research suggests that managers view strategic issues as threats unless there is strong evidence to do otherwise. Thus, the advice to view all issues as entailing opportunities (as would be natural for proactive managers) may be far easier to give than to follow.

The Leadership Challenge

Leadership teams navigating digitally enhanced turbulence must confront the sticky nature of organizational attention—and reorient people's attention through dialogue, task assignments, incentives, training, and *investments* in foresight activities. Here are four principles to guide a team:

1. *Use available digital technologies to measure where organizational attention is high and low.* For example, by analyzing work related emails in a firm, suitably anonymized, leaders can track what issues are trending. Such text-analytic approaches are used widely to assess consumer sentiment in the travel industry or for early detection of shifts in the appeal of political candidates. Sentiment-analysis software can handle massive amounts of data, from everything published in popular media about a personal matter to emails, intranet content, or other corporate communication formats. Many will recoil, however, against this kind of intrusion by Big Brother, even if the scanning is anonymous. Significant concerns are growing about protecting privacy when using digital technologies, as reflected in Europe's new

General Data Protection Legislation.[28] The legal rights and reasonable economic concerns of companies clearly need to be better balanced against individual rights and social norms.[29]

2. *Recognize that prior knowledge shapes the creation of new knowledge inside a firm.* New information can only create value if it connects with existing know-how. The richer a firm's existing knowledge base, the finer will be its sieve for catching new information about a topic. If leaders feel more attention should be paid to customer service, regulatory compliance, or some new promising technology, then they need to train people in those domains. This in turn will enhance the firm's absorptive capacity in those areas and draw organizational attention there. As Louis Pasteur noted, chance favors the prepared mind, and various techniques—such as scenario planning, scanning exercises, and war gaming—can help prepare the corporate mind to get luckier or smarter.
3. *Although focused attention is crucial to understanding new information, too much of it can backfire.* Focusing intently on one area comes at the price of narrowing the peripheral vision of things happening elsewhere. To avoid running through red lights, leaders must create slack to explore beyond the firm's field of vision. Leaders should encourage curiosity about relevant topics that may seem removed from present concerns. They can create task forces that counter the prevailing focus areas of the organization, such as forming a "red team" to challenge whether a new strategy really is going to succeed.[30] The red team plays the role of the loyal opposition, periodically reviewing the assumptions and progress of the blue team, which is tasked with executing the strategy. The benefit is a faster way of collecting information and tracking progress, with sufficient time to make midstream adjustments when needed or to terminate the project entirely.
4. *Encourage managers to develop a third eye to help notice hidden cues or soft signals that matter.* When meeting with customers or external partners, leaders should pay as much attention to what is *not* being

said or what is being hinted at between the lines. When examining the murder of a horse trainer, fictional detective Sherlock Holmes embraced this principle by asking a local constable about the curious incident of a dog not barking in the night. Holmes deduced from this missing cue that the dog knew the murdered person. A more disastrous example of not spotting missing data occurred in 1986 when NASA examined a data chart of previous shuttle flights the night before the scheduled launch of the Challenger. The concern was that low temperatures could cause O-rings to fail, but NASA's chart showed no correlation between past O-ring damage and ambient temperature. Sadly, the chart did not include flights with zero O-ring damage, which would have clearly established a link. NASA proceeded with the launch, and a few minutes later the shuttle exploded in midair, killing all aboard.

It's often hard to notice what may be missing when solving problems in real life, but training in statistical reasoning, being aware of confirmation bias, and examining past cases all can help.[31] The NASA case, for example, concerns the common error of looking at just a subset of the available data, akin to doctors diagnosing a new illness by looking only at sick patients without comparisons to similar patients who are still healthy. Recognizing this classic error is now part of NASA's own training, as well as many executive education programs. Deeper awareness about our own decision biases due to cognitive, emotional, and social factors is critical in managing attention in an organization, which is the subject of the next chapter.

4 Sensing Weak Signals Sooner

> Judge a man by his questions rather than his answers.
>
> —Voltaire[1]

Modern-day decision-making is complicated by a relentless decline in the ratio of useful signals to distracting noise, leading to pervasive information overload, diversion, and confusion. At the extreme, a leadership team can be paralyzed into inaction, burying its collective head in the sand. The fact is that all of us will miss important signals at some point. Activating our sensing capability requires two decisions: *where* to look (*scoping*) and *how* to look (*scanning*). The challenge is to flexibly balance a narrow focus with sufficient scanning of the changing periphery around us, and there is no single right way to do this. What's required for a hospital system trying to understand new digital business models will be very different from the approach of a global confectionary company. Nonetheless, we have found that the deeper principles of sound sensing apply broadly.

It Starts with Better Scoping

Most industries today are prone to signal overload and confusion, compounded by the disruptive effects of new technologies. Even the seemingly stable chocolate confectionary business is not immune. Today, $75 billion per year is spent globally on chocolate treats. Yet many of the big players, including Mars, Hershey, Cadbury, and Nestlé, have missed relevant signals. One European chocolate maker was blindsided by the merger of Mars and Wrigley; another failed to develop new marketing campaigns in response to changes in consumer behavior around *how consumers eat* and *what they buy*; several were late in adopting digital media to monitor product flows and changes in consumer buying patterns and using that data to update their packaging; and others failed to segment their market to develop new products geared toward people with different lifestyles.

To illustrate the diversity of issues that a leadership team must consider in this particular business, figure 4.1 shows a sampling of the uncertainties that lurk on the periphery of mass-market chocolate production.[2] They span from technological innovations to environmental

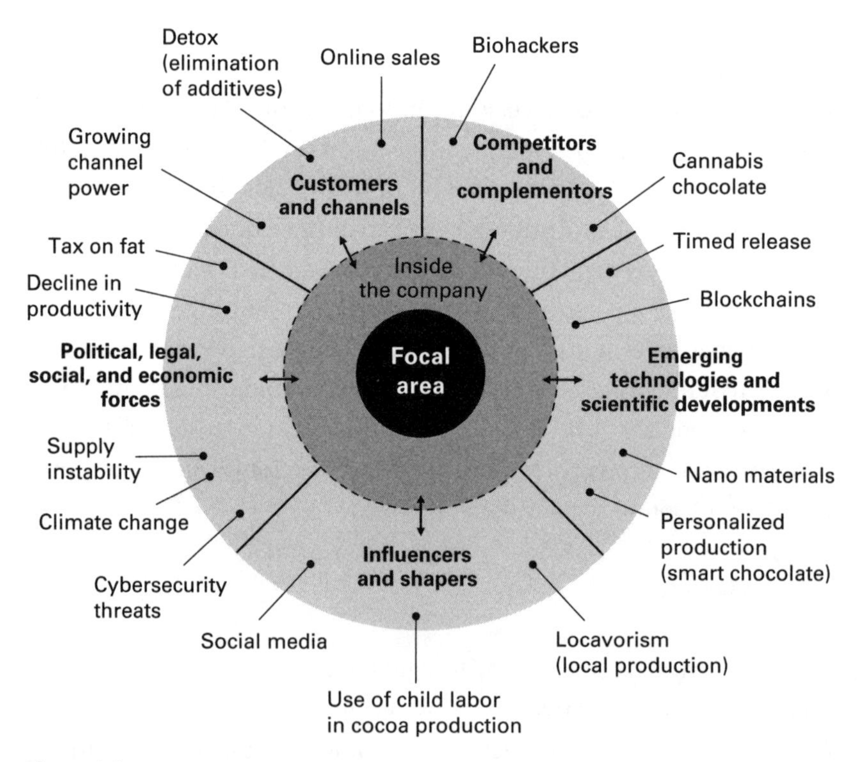

Figure 4.1
Scanning the external periphery

or political issues. For example, Turkey is a major global producer of hazelnuts—a popular addition to chocolate bars—but its supply source is at risk due to internal unrest and Turkey's geographic proximity to the violence in Syria and Iraq.

Each of the five zones in boldface in figure 4.1 continuously emits weak signals that a chocolate maker needs to track—while remaining alert to any worrying signs from within the organization itself. In an era of digital generation and amplification of weak signals, each source can spawn signals that soon overwhelm an organization's ability to absorb them.

The art of scoping entails deciding which sources truly deserve closer attention. Choosing the right scope has less to do with knowledge than with curiosity and relevance. It hinges less on knowing the answers than on posing the right questions to reveal the limits of current knowledge. It also means sensitizing the organization to issues that are still under the radar. In this spirit, Google cofounder Larry Page has challenged his teams to anticipate the future not just by asking what is or likely will be true, but also, what *could perhaps* be true, even if totally unexpected.[3] Such guiding questions are a productive way to launch a scoping dialogue.

Most managers are very proficient at asking and answering focal questions about things such as departures from budget, weekly sales volume, and cost variances. These questions are often so routinized that they are displayed in a dashboard of metrics and watched closely. *Guiding* questions, on the other hand, are meant to be thought starters for probing the parts of the picture that likely are missed by focal questions. These questions can be organized into three categories: learning from the past, interrogating the present, and anticipating the future.

Questions That Help You Learn from the Past

The past may not be a good predictor of the future, but it can reveal persistent blind spots. One of the most compelling guiding questions is, "What have our past blind spots been?" Recall from chapter 2 that a diagnostics company recognized its systemic blind spots by cataloging the threats and opportunities it had seen early versus too late. After surfacing these misses, we worked with the leadership team to find patterns to explain them, thus revealing systematic vulnerabilities. We then expanded this diagnostic inquiry to include internal threats *and* opportunities.

Useful lessons can also be gleaned by asking, "Is there an instructive analogy or precursor from another industry or geography?" Armed with this question, one nanotechnology firm began looking carefully at the genetically modified organisms (GMO) controversy in Europe (remember the Frankenfoods scare?) for possible indicators of public resistance.[4] GMOs and nanotechnologies have some worrying similarities: both have been viewed as presenting health hazards; both were developed by faceless, global firms, whose motives are often regarded with suspicion; and in both cases, the public could easily imagine various hazards, while the supposed benefits were mostly indirect. Such analogies enable leaders to see their own situations through a wider lens.

In the same way, precursors from other markets can be a wake-up call or an early signal of an opportunity. For example, one vigilant packaging technology firm keeps an outpost in markets like Japan, where innovations may first appear. An *outpost* is a person or group that serves as the eyes and ears of the parent company to help it see interesting developments sooner. Procter and Gamble (P&G), for example, keeps some retired executives in Europe on a part-time retainer so that they can periodically report on interesting developments in, say, private label or branded products. For P&G, this is part of its "connect and develop" approach to innovation, which helped it launch a highly successful rotating toothbrush years ago.[5]

Other insightful guiding questions include, "Who in our industry has a consistent record of seeing sooner and acting faster?" and "What is their secret?" These questions may reveal best practices that can be adopted to improve your own vigilance. This stage of questioning is at best a starting point to reduce your vulnerability. But improving vigilance requires a lot more than dwelling on the past.

Questions That Interrogate the Present

Sometimes signals are right in front of a leadership team but are barely noticed or appreciated. People have a powerful tendency to filter out warning signs and pretend that all is well.[6] Indeed, the smarter we are, the better we become at rationalizing away signals of impending problems. We have found that most surprises have antecedents, and they often appear as anomalies—something that deviates from what is normal or expected. Perhaps a competitor is hiring a different kind of talent, a customer complains about a tightening of labor supply, or a trade rumor emerges about price cutting. A focus on such anomalies was central to Alan Mulally's efforts to challenge the insular and vulnerable mindset of the Ford leadership team, as we described in chapter 1.

Most organizations have some people who truly think outside the box, and much can be learned from these mavericks and outliers. However, Andy Grove, the famed proponent of *healthy paranoia*, found that most mavericks within Intel had difficulty explaining their visceral concerns and feelings to leadership;[7] as a result, their insights were usually ignored. In addition to encouraging mavericks to speak up, vigilant organizations also listen carefully to employees at multiple levels and encourage openness to different viewpoints. This helps them spot opportunities and catch festering problems before they explode. These efforts go far beyond suggestion boxes; they include regular meetings, formal recognition, and feedback on what happened when a reported signal was considered.

Another rich source of guiding questions can be unsatisfied customers—especially those that defected to another firm. Lost sales reports and postmortems on contracts won by competitors can be revealing—but only if those doing the pathology are open to digging deeply and sharing their learning. Most firms monitor blogs, social media sites, and chat rooms for signs of trouble to take rapid, remedial action—but some go further to surface anomalies, exceptions, dissatisfactions, and open-ended complaints. P&G did this after launching Febreze, a spray that eliminates unpleasant household and fabric odors. Rumors started in chat rooms that Febreze might kill canaries or other birds kept as pets. Because P&G is so sensitive to and alert about possible damage to its brands, these rumors were quickly picked up, independently investigated, and then successfully countered by the company.[8]

Vigilant firms try to be especially alert to changes in the behaviors and needs of their customers. A powerful way to canvass the present for clues about the future is to look for *edge cases*[9]—early indications of trends that *could* become major opportunities or threats. One company asked, for example, "What are some new jobs that didn't exist before 2018?" The answers included indoor farmer, synthetic tissue engineer, and virtual fashion designer. The team could roughly infer the nature of these job postings and determine whether a trend watch was merited. But other jobs that surfaced, such as bot wranglers, were beyond most managers' experience base and would require deeper investigation.

Questions That Anticipate the Future

Guiding questions about the long term can be greatly aided by the construction of scenarios. These are alternative future narratives that reflect the uncertainties in the environment today and how they may play out in years ahead.[10] This methodology aims to magnify potentially important weak signals by providing a broader context that makes them more salient and connected. If different scenarios highlight a particular weak signal, the organization is less likely to filter it out. To see the full impact of possible future surprises, a leadership team should ensure that there is at least one unthinkable—or at least hard to believe—scenario. This is a possible future deemed *so unlikely* by the majority of managers that it isn't getting any serious attention—for example, the demise of your firm within the next ten years, as happened with Arthur Andersen and Enron. By explicitly entertaining unthinkable possibilities such as key leaders going to prison (Enron) or the business model becoming obsolete (Kodak or Nokia), a team is forced to recognize the diverse ways that signals can be interpreted. Without such shock interventions, the collective mind is prone to force-fit any faint stirrings into the prevailing mental model and avoid uncomfortable or taboo subjects.[11]

Another thought starter is for leadership team members to ask how they would attack their own business if they were part of a new market entrant deploying a disruptive digital business model. The aim is to surface and exploit all internally known weaknesses of the current business. This can be done by setting up an internal team or by bringing in outsiders for fresh perspectives. This type of thought experiment sensitizes everyone to look for indicators that such an attack is in fact being mounted, creating a highly prepared collective corporate mind.

This approach is a variant of the "red team" exercise used by the military, mentioned in the previous chapter, whereby internal teams play the roles of key decision-makers in a fictitious competing firm.[12] Within this simulation, they can respond more creatively to possible strategy changes or speculate on moves they would make as a competitor. Red teams can also be used to detect weaknesses in a new strategic initiative such as launching a new product or integrating a merger. In each case, the team would be asked to systematically collect any signals that the plan as designed may not succeed—thus allowing for timely corrective action.

Imaginative guiding questions about the future might include the following: What surprises could really hurt (or help) us? Might there be future surprises of the same magnitude of impact as those that have already occurred in recent decades? For example, in financial services, what future surprises might be as big as PayPal or Apple Pay or the various regulations imposed after the Great Recession?

Leaders can also conceive an idealized future and then work backward to envision the changes that would have to occur to realize it. Russell Ackoff, the late, renowned professor at the Wharton School who became known as the dean of systems thinking, called this *idealized design*. In this approach, a leadership team is asked to create an organizational design based on a visionary distant point in the future, ignoring current obstacles or constraints.[13] For example, a global advertising agency we worked with was asked to imagine a marketing department of the future that would fully leverage the emerging array of digital technologies. How would the role of the chief marketing officer or chief commercial officer change? Where and how would media decisions be made? How would the marketing group orchestrate a group of partners to implement these decisions? Such questions prompt deeper queries that prepare leaders for early alerts of unexpected changes.

Telescope versus Microscope

In some cases, the right tool will be a telescope to look over the broad horizon; at other times, a microscope is better to examine a small part of the terrain in great detail. Once a vigilant leadership team is clear about its guiding questions, it becomes easier to pick the right scope. The deeper analyses this entails can be supported by third parties, like consultants, as an added responsibility for the marketing or strategy group, and/or by forming a special task force for complex issues. For example, General Electric's senior healthcare group created a task force to process an array of weak signals about new opportunities for healthcare in India.[14] These signals suggested several nonlinear shifts, including a shortage of doctors and hospital beds, growing unmet healthcare needs, and an underdeveloped health insurance industry, but also good digital connectivity. The mandate for the task force was to illuminate areas that might deserve more attention as interesting potential opportunities for GE in India.

The choice of scope will evolve through an ongoing process for channeling the priorities and curiosities of the organization. As a result, the questions in each category must be continually reviewed, refreshed, and updated with new insights gleaned from *scanning* processes. In our experience, most managers scan passively, keeping their antennas up to receive signals from a range of sources directly related to their industry—from trade rumors to technology forecasts and performance metrics. The danger of this approach is that the data mostly comes from familiar sources, can be overwhelming in volume, and is largely interpreted to reinforce rather than challenge prevailing beliefs.

Active scanning, on the other hand, reflects leaders' curiosity about bigger uncertainties and is driven by creative hypotheses that shed new light on the preliminary evidence collected thus far. For example, an auto insurance company found that its clients were dropping their contracts more often than in the past but returning as customers later. Apparently, they were not dissatisfied with the company's pricing policies; what other explanation could account for this behavior? One disconcerting hypothesis was that the market was becoming "frictionless," with low switching costs to explore alternative providers. A more promising hypothesis was that there were opportunities to sell time-limited, activity-oriented policies that enabled customers to switch policies easily. This hypothesis was then tested through some further exploration, which validated the conjecture.

In its optimal form, active scanning approaches the scientific method, which starts with multiple hypotheses based on current data. Additional targeted data then is generated, followed by keen observation, resulting in further speculation and tests until deeper insights have been garnered that are actionable for leaders. The next four subsections discuss principles of a strong scanning practice.

Encourage Divergent Thinking

Constructive debate and divergent thinking yield deeper insights[15] and help to counter the problem of widely distributed intelligence that plagues most organizations. Doing this well means actively encouraging diverse—and even contradictory—inputs to ensure that all sides of an issue are surfaced.[16] When a leadership team is surprised by some circumstance or an unexpected event, usually there were people inside the company who knew about it but didn't raise their voices. To promote diversity and encourage contrarian views from all parts of the organization, the CEO of Arby's, a fast-food chain with over 3,300 locations, deliberately surrounds herself with colleagues of different races, geographies, and personality styles. "You really don't need another you," she says.

The opposite of encouraging diverse thinking is for leaders to coalesce around the seductive comfort of shared viewpoints and succumb to groupthink. Amazon's Jeff Bezos decries such social cohesion as the "cloying tendency of people who like to agree with each other and find consensus comfortable." Instead, Bezos fosters a culture in which leaders can challenge perspectives they disagree with, "even when doing so is uncomfortable or exhausting."[17] Ray Dalio, the founder and CEO of Bridgewater, one of the world's most successful hedge funds, embraces this philosophy to the extreme. He even enshrined it in his book, *Principles: Life and Work*, putting radical transparency at the very top of his list of priorities. This is easier said than done, however, because the benefits of brutally honest debate can dissipate quickly if some participants find it uncomfortable or feel offended. Such deep dialog requires mutual trust and respect and a "meeting of the hearts" before a genuine meeting of the minds can occur. It may entail bridging differences by creating engagements that build trust.[18]

Think from the Outside In

Vigilant firms are better at stepping outside the boundaries of their firm to look at their market through the eyes of customers and competitors and ask key questions: How are our customers changing? What new needs will they have? How will they satisfy them? And who will we be competing with in the future? The logic of starting with an expansive view of market possibilities and then seeking deep market insights to test hypotheses about these possibilities is compelling. Yet it's often resisted in favor of starting from the inside on the grounds that the sales people are always closest to the market, R&D knows all the technological possibilities, and leadership understands the business best. Steve Jobs felt that "a lot of times, people don't know what they want until you show it to them."[19] Although Jobs may not have been guided by formal market research, he was intently focused on the complete user experience and backed up his intuition with an enormously expensive design process that continuously sought feedback from users. Also, he was a wonderful counterweight to the industry's prevailing engineering culture, which values technical wizardry over ease of use, aesthetic thinking, and beautiful or smart design.

The danger of inside-out thinking lies in the premise that resources exist to be used and that the task of management is to exploit them fully.

Superior resources such as excellent service operations and a strong supply chain create advantages that take time to build, but they also limit the ability of the firm to adapt. By focusing on internal resources and skills as a starting point, the strategy dialogue is prematurely narrowed and anchored. This problem can be exacerbated by using SWOT analysis, which focuses on understanding current *strengths* and *weaknesses* as well as *opportunities* and *threats*. As noted in chapter 3, the very use of such labels as *threat* or *opportunity*—which are common concepts in business speak and hard to banish as such—implicitly assumes the status quo as the relevant reference point. This may indeed be a valid benchmark, but an outside-in view would favor more neutral labels such as *issues or developments that deserve attention*. This avoids prejudging these factors as either good or bad a priori and getting trapped in a self-fulfilling prophecy about threats especially.

Amplify Interesting Signals

Active scanning tends to surface far more signals of looming threats or potential opportunities than an organization can possibly digest, resulting in a paucity of attention (as discussed in chapter 3). The initial divergent thinking phase therefore must be counterbalanced by systematically amplifying, clarifying, and then converging on the most interesting signals. There are five robust ways to aid the process of sifting through the noise to reveal signals in need of further attention:

Canvass the wisdom of the crowd. Studies show that large groups or crowds often make more accurate collective judgments than individual experts.[20] The basic idea is that groups can be collectively smarter than even the smartest people in them. One reason is that individuals in an organization have only partial information and are thus prone to error. So long as this error is randomly distributed across the group, taking an average largely cancels out the noise and better reveals the signal. However, this doesn't work if people's errors are correlated—as can happen if groupthink or some other collective blind spot sets in, pointing everyone in the wrong direction. The wisest crowds are those that are the most diverse, such that people's views are not too positively correlated. To get the full benefits, try adding people whose experience, contacts, and perspectives are actually *negatively* correlated with the group as a whole to which they're being added. If the crowd is digitally connected via chatrooms, emails, or video conferences, an organization can gather everything from dispersed intelligence about internal innovation opportunities to judgments about competitive moves or the threat of an emerging technology platform. Importantly, the potential of group wisdom is fully realized only when good guiding questions are asked. Also, leadership must be completely supportive—at which point the crowd can become an efficient intellectual marketplace.

Leverage the extended network. All organizations are embedded within networks connecting them to partners, suppliers, distributors, researchers, and consultants. These network relationships offer the possibility of extending the reach of their sensing systems—albeit at the risk of further overloading the firm's capacity to absorb weak signals.[21] Increasingly, organizations are adopting an open-systems approach that greatly expands the number of nodes of connection to their ecosystem.

One exemplar is Li & Fung, which generates over $8 billion per year in sales of garments, toys, and other assembled products. Similarly, Apple connects to over a million software developers, thousands of accessory makers, and a myriad of content suppliers, each with a sensor capable of picking up weak signals.[22] For example, a supply chain can be repurposed to gather intelligence about industry dynamics, competitor behavior, or impending supply shortages. We'll discuss later how to absorb and interpret the plethora of weak signals with an organizational strategic radar that combines scenario planning, business analytics, and dashboard technologies for continuous monitoring. The goal is to harness the power of digital technologies to digest the weak signals they collectively produce.

Apply successive filters. How should the pool of possible growth opportunities surfaced through an active scan be pruned to converge on the best bets? This is the same filtering question to be asked of the weak signals of threats and opportunities from all domains. The difference here is that the methodologies for screening opportunities are further advanced, applying successively tighter filters that are generous at first (when uncertainty is high but relatively few resources are at stake) and more restrictive later (when the situation is reversed).

The first filter involves running an "innovation tournament" to define the opportunity set, using informed insiders to refine and evaluate a roster of growth opportunities.[23] The goal is to reach deep and wide into the organization for ideas that would otherwise remain dormant. Participants are invited to suggest whatever they want—including crazy ideas: the more ideas enter the innovation funnel, the greater the chance that something of value will emerge at the other end. In addition to scoring and filtering all the bottom-up ideas, teams can also dissect, debate, and recombine ideas to make them stronger along the way. Innovation tournaments can be conducted online or in a one- or two-day workshop to encourage intense dialogue. Contrary to popular wisdom, individuals working alone can often be more productive at generating innovative opportunities than those same people brainstorming in a group setting. Why? Because of the bottleneck created when people speak one at a time and group dynamics such as groupthink, which inhibit the creative flow of ideas.[24]

The next filter takes the refined ideas from the innovation tournament and feeds them one at a time through a comprehensive screening tool[25]—a disciplined framework by which teams identify faulty assumptions, gaps in knowledge, and potential risks. This is usually deployed as a succession of scoring screens incorporating increasingly detailed and ever more accurate answers to key aspects of an evolving opportunity.[26] Natural questions include: Has this been tried before elsewhere? What were the implementation challenges, cost, and benefits along the way? How long will a pilot test take? Who might champion the idea? Cumulative learning about an opportunity takes considerable investment to avoid even larger and possibly irreversible investments that could be lost in the ill-conceived pursuit of an opportunity.

Triangulate Multiple Perspectives

Renaissance artist and inventor Leonardo da Vinci emphasized the virtue of looking at things from at least three different points of view. Just as a GPS uses three coordinates to place you on a map, managers should use multiple enquiry methods to clarify ambiguous signals and then probe deeply to learn more about promising patterns. One way to embrace this approach is to elicit viewpoints from external experts or advisors, such as the Institute for the Future—an organization that offers structured workshops to help leaders of firms as diverse as eBay and Lufthansa triangulate weak signals and envision different possible futures. This Silicon Valley–based institute also convenes communities of experts, such as the Blockchain Futures Lab to examine what a world driven by blockchains might look like a decade from now.

No single method can do it all. Since any given method has its own distinct limitations, leaders need to explore multiple approaches. Managers often use previous experiences or cases to understand how an emerging technology such as blockchain might evolve. But these past analogies seldom capture the full picture because the situations are not fully comparable. For example, some people may think of innovation as analogous to cultivating a garden since both need seeding, nutrients, pruning, and pest control. Others may see innovation more like a game of poker since both involve money, information asymmetry, strategy (including bluff perhaps), and knowing when to fold is crucial in a zero-sum game. These two metaphors (of gardening and poker) cover many aspects of innovation but still fail to capture other relevant aspects such as the role of invention and team work or the influence of a referee and changes in the rules.

So let's apply the need for multiple approaches to the future of blockchains, which "combine principles of law, math, game theory, cryptography, monetary policy and computer science into an open source software platform."[27] A blockchain-based platform differs from an Internet-based one in that blockchains allow "one person to send a bitcoin or other cryptographically secure digital asset to only one other person," which is a major difference.[28] The Internet is essentially a virtual copying machine that creates digital abundance; in contract, a blockchain is about creating digital scarcity. Rather than copying your manuscript file and sharing it simultaneously with many others, a blockchain allows you "to send each person an access token linked to a specific manuscript file. Only the token holder can access the associated manuscript version. The holder may transfer the token and you can track it from owner to owner."[29]

Still, it is understandable that people view the Internet as a natural analogy to explore the future of blockchains since they both benefit from network effects.[30] In addition, blockchains allow people who don't necessarily trust each other to collaborate in a relatively safe, secure, and autonomous manner, much more so than the Internet permits. So, the Internet analogy has various limits in projecting the future of blockchains and other analogies will be needed, such as virtual reality perhaps or biological ones such as gene editing. The intent is to surface other distinctive features of blockchains, in addition to key differences, such that various analogies combined add more insight into the future of this new technology. Such deeper understanding will help envision new applications of blockchains in other domains such as publishing. Thanks to its unique identification of who received, modified, or posted a document when, blockchain will allow far better digital rights management solutions in publishing, patents, journalism, art, and other domains yet to be imagined.

Explore diverse angles. When triangulating, it's key to surface viewpoints that are sufficiently diverse—perhaps even opposing. As with the wisdom-of-the-crowd approach, the power lies in aggregating across diverse viewpoints to eliminate random noise and amplify whatever wisdom exists. Prediction tournaments—in which competing teams develop their own forecasting methods—are another expert aggregation approach.[31] Each team is asked to make subjective probability judgments about some key events, which then can be either averaged or weighted differentially depending on track records and confidence levels.[32] Even if there is high expert consensus about the odds of some event happening, the group's mean guess may not be the best estimate.[33]

Exploring ambiguities is about seeing complex issues from diverse angles. Close-knit groups inside an organization may offer each other reassurances, but this can also lead to dysfunctional groupthink. Research shows that less rigid groups involving outsiders can often lift a team's IQ considerably in many domains.[34] This is particularly true when companies use feedback mechanisms (such as Delphi polling) to tap into the collective wisdom of the organization. The Delphi technique starts with experts giving their individual opinions (anonymously or with attribution). Each viewpoint is then debated by the group and the experts provide another round of predictions or estimates. This continues until no person makes any further adjustments, at which time a (weighted) average is taken.[35]

Creating anonymous "opinion markets" is another way to avoid collective myopia. For example, Hewlett-Packard asked employees to participate in a newly created opinion market to forecast sales. Employees could bet in this market over lunch or in the evening at home, revealing through their wagers where they felt the market was headed. *The result:* These forecasts beat the traditional company forecasts 75 percent of the time.

Catch smoldering crises. Vigilant organizations not only scope and scan their external environment, they also actively seek signals from within their own internal environment—and for good reason. The Institute for Crisis Management (ICM) tracks negative news in business that triggers adverse consequences such as accidents, labor disputes, workplace violence, discrimination, sexual harassment, and more. Many news items involve internal threats that management spotted too late. Indeed, mismanagement was the major cause of crises that made the news, accounting for 27 percent of the total in 2017. "Smoldering crises," which likely had early warning signs, made up 71 percent of the cases reported in the news that year. These can be very costly to organizations, as we saw in chapter 1, which featured egregious examples such as Volkswagen, and Wells Fargo Bank.

Echoing our advice, the ICM recommends examining past cases, exploring your organization's vulnerability to future crises and learning from the mistakes of rivals and peers. It also suggests working to create a "goodwill bank" with key stakeholder groups ahead of a crisis through clear communication, transparency, and honesty because your firm might be featured negatively in the next viral video on social media. Building goodwill with stakeholders requires diverse employee engagements, including a plan for when, where, and how to stay in touch with key allies—as well as critics, perhaps. Having developed relationships with media producers, community leaders, analysts, and opinion polling experts can help when a crisis strikes.[36] The key in all of this is clear and honest communications, admission of errors when warranted, and the avoidance of evasion or cover-ups.

Parting Guidance

Leaders are not typically judged on the quality of their questions, nor do we design our educational systems or executive development programs to develop this critical capability. Yet when an environment is changing fast and uncertainty is pervasive, the prizes will go to those asking the best questions. Inquisitive leaders, however, need to carefully balance their desire to get key guiding questions answered with the limited resources available. Four general principles can ensure that the sensing and exploration process is balanced and practical:

- In most cases, the problem is not a lack of data, but a lack of good questions. Managers may console themselves by gathering more information, but unless they widen their field of vision, they may not see important opportunities and threats soon enough.
- Sense actively, with an open mind and a mix of directed hypotheses and exploratory journeys into the unknown. Active scoping and scanning is not a one-time or episodic event. It must be a continuous learning process that employs a broad repertoire of approaches and engages the entire organization.
- When faced with weak signals or ambiguous data, resist the urge to jump to conclusions; instead, generate multiple hypotheses and

explore further. As Voltaire observed, "Doubt is not a pleasant condition, but certainty is an absurd one." It takes courageous leaders to admit that they don't have all the answers and then lead their team into new avenues of exploration—and, eventually, deeper insights.

- Commit the organization to develop sensing as an important strategic capability. This starts with those at the top behaving according to the three qualities of a vigilant leadership team introduced in chapter 2: foster curiosity, apply strategic foresight, and nourish a culture of discovery.

The relevance of any weak signal—whether a faint warning or an ambiguous alert—will seldom be clear on its own because it depends on context and broader circumstances. By now, for example, most firms are alert to the possibilities of AI due to advances in machine learning, deep neural networks, and an exponential increase in the availability of digital data. The implications of this inflection point in AI are quite different, however, for an appliance maker designing smart kitchen devices than for a ride-sharing service seeking a real-time connection with its customers. To understand such important contextual differences first requires judicious sensing and then timely, as well as flexible, decision-making, as we'll discuss in the next chapter.

5 Tackling Ambiguity

> Writing is like driving in a car at night. You can only see as far as your headlights, but you can make the whole trip that way.
>
> —E. L. Doctorow[1]

For today's business leaders, acting on weak signals is much like Doctorow's image of driving at night. The destinations hinted at by emergent signals are usually obscured by darkness; to avoid going into the ditch, you must proceed cautiously. This is even truer in a world of digital transformation. The myriad paths digital technologies can take are compounded by the unpredictable actions of regulators, competitors, and customers. This chapter follows the steps taken by vigilant firms like Philips Lighting to illustrate the challenges and the promise of this journey to action. Figure 5.1 shows how this chapter bridges the sensing activities of chapter 4 and the timely action taking that we'll describe in chapter 6.

Shining Light into Darkness

The emergence of light-emitting diode (LED) technology as a potentially disruptive market force began faintly in 1962, when an MIT researcher coaxed a semiconductor to glow in the infrared range. As LED technology gradually improved, the aim became to produce a white light, ideally as bright and warm as that found in homes and offices. This

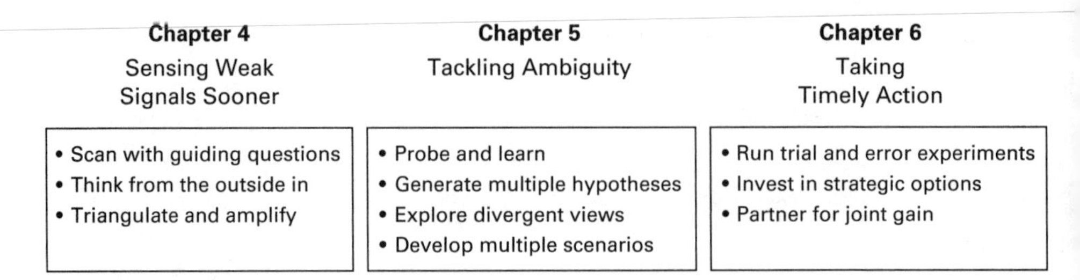

Figure 5.1
From scanning to seizing

development posed a serious threat to the lighting industry, which was still reliant on centuries-old incandescent technology.

As a result of commoditization and intense price competition, the lighting market was shrinking, which distracted leaders from attending to the LED threat. Most managers in incumbent firms remained intently focused on short-term profit targets in their core markets, while novel players specializing in efficient LED lighting entered the marketplace with innovative, efficient light displays and clever lighting solutions for small gadgets. The biggest wake-up call came with the loss of the US traffic light market to LEDs—a substitution that allowed cities to use 90 percent less energy and with far longer lighting lifespans. By 2003, *MIT Technology Review* had put incandescent lighting on its endangered list of the "top ten technologies that deserve to die."[2]

When an organization detects a weak signal, it's seldom obvious whether it should be put aside or pursued further. Four tools are especially useful in reducing this ambiguity and interpreting the implications. This chapter shows how the incumbents in the lighting industry could have used these tools to understand how to respond to the LED challenge.

Probe and Learn

This starts with unanchoring yourself from conventional wisdom and myopic views. As Voltaire counseled, we should not judge people by their answers but by their questions because questions reveal more about how well and creatively people think. How proficient is your leadership team, for example, at placing itself in the shoes of key stakeholders? This includes asking bold questions that challenge the status quo. As noted earlier, Andy Grove did so to change Intel's core strategy away from memory storage and toward semiconductors. The ability to challenge your own and other people's views calls for three qualities: (1) humility in accepting that no one person can see all angles, (2) courage in flinging the window open to whoever can offer new insights, and (3) developing a tolerance for well-intentioned wrong decisions or judgments.

When assessing a potential digital disruption, leaders can begin by probing the latent needs of current as well as potential customers more deeply. End users may not be able to tell you how to design a product or service, but customers are often articulate when asked about their problems, changing needs, and circumstances, as well as the aggravations they encounter during actual usage—or their "pain points." These answers often reflect latent needs as opposed to manifest, clearly evident needs. Various tools are available to probe and learn more about latent needs, including problem-identification workshops, customer experience mapping, and metaphor-elicitation methods to better understand consumers' psychographics.[3]

Other ways of hearing the voice of the customer to surface possible opportunities include the following:

- Leveraging lead users. These are customers facing needs in advance of the rest of the market for which we are working to find solutions sooner.
- Monitoring complainers and defectors for clues as to why their needs were not met and how they are solving their problems.
- Hunting for precursors in other regions or countries, where fads, fashions, or digital innovations tend to appear earlier.

Generate Multiple Hypotheses

Vigilant organizations are superior at generating mind-expanding hypotheses about the deeper meaning of the weak signals they need to probe further.[4] The aim is to explain anomalous events, such as the deeper threats from an emerging technology like LEDs or the intentions and strategies of new entrants. The initial hypothesis may not be correct, but it jump-starts a vital testing process in a disciplined way until better insights emerge. Such creative testing usually generates even better hypotheses, focused on yet bigger issues of concern, thus setting in motion an iterative and cumulative learning process.

Given the power of groupthink, formulating competing hypotheses is not easy. The key is to recognize that conflict can be constructive, especially when focused on tasks rather than relationships or personalities. The extremes of either no conflict or maximum disagreement seldom work well. Studies confirm that *moderate* conflict leads to superior decisions because it pushes team members to gather better intelligence, explore more options, and examine issues in more depth. In contrast, more harmonious teams may miss key pieces of the puzzle, and extreme ones get stuck or frozen in polarized views.

For example, a maker of a technical equipment used at many remote customer sites was struggling with poor on-site maintenance, which became costly and left its clients angry.[5] By this point, the signal was pretty strong—but what should be done about it? One of the hypotheses proposed was this: If customers can easily get us better information about their problem, we'll be more likely to bring the proper equipment. To test this hypothesis, the company ran an experiment to see whether customers would email detailed photos of their problems. It worked, and this simple example shows how creative hypothesis generation can help resolve ambiguity.

In the LED case circa 2000, the leadership of incumbent firms needed to develop multiple hypotheses about several looming issues:

- *Development of technology:* Industry leaders needed to assess the pace of development of high-quality white light and major reductions in cost. In 2000, the projections of future prices for solid-state technology ranged from fourteen dollars per kilolumen to as low as fifty cents. Likewise, efficiency estimates covered a wide spectrum, from four to eight times that of incandescent lighting.[6] These very broad ranges implied that the new technology could capture anywhere from 10 to 90 percent of the current market for incandescent and fluorescent lights. Clearly, lighting companies needed to develop sharper hypotheses about the pace of cost reductions and increases in efficiency to get more refined estimates. But companies with big legacy investments in the established supply chain were slow to act and myopic about the possibilities of digital lighting. This "industry lethargy" allowed a few smaller entrepreneurial firms to establish themselves in the emerging digital lighting space. Most incumbents eventually saw these small to midsize firms as potential disruptors but were unclear what to do. Philips Lighting was among the first to respond by partnering with the upstarts while investing internally in technology development. The rest of the lighting industry was slower to act, which allowed Philips to establish itself as the clear leader in the emerging digital lighting ecosystem.

- *Changes in the attitudes of trade and end users:* Although the technology itself could facilitate LED adoption, the willingness of contractors and end users to go along with it could have been a limiting factor in the adoption of LEDs. About one-third of the total lumen-hour demands were replaced and/or installed per year, which limited the maximum penetration rate of this new technology. Managers therefore needed to develop strategies for addressing both contractors and end users, as well as intermediaries such as designers and specifiers, who influence consumers' lighting decisions. This called for new hypotheses about market and channel developments. As with most emerging technologies, LED faced an uphill battle getting a foothold in a well-established value chain. Entrants tried to disrupt the industry, but the incumbents dug their heels in deeper—trying to slow down the adoption of LED technology. A few pioneers in the industry, however, delved more deeply into the promise of the new technology, while other key players in the value chain (such OEMs, distributors, and lighting designers) watched with skepticism. Some questioned the long life expectancy touted for this new light source, whereas others were concerned about the impact that the LED technology's much longer projected life span would have on their replacement sales.

- *Lobbying and public affairs:* Attitudes toward conservation and energy consumption were also evolving, as reflected in both market perceptions of the technology and government regulations. One forecast was that the widespread adoption of LEDs could reduce the worldwide consumption of electricity for lighting by more than 50 percent, resulting in a 10 percent decrease in electricity consumption globally. Managers therefore needed to generate competing hypotheses about how to interact with multiple government agencies and with the media through lobbying and public affairs. The role of nongovernmental organizations (NGOs) and government agencies proved significant in accelerating the market impact of LEDs. Especially in the United States and Europe, there were significant efforts to support LED innovation by offering incentives for technology development. Few legacy companies fully understood the role this played in their own sandboxes. In particular, these public-private partnerships boosted the penetration of LEDs in traffic signals and street lighting, as we'll discuss later.
- *Wild cards:* Further hypotheses about the relationship between lighting and health were also needed and evolved slowly as well. Research had already shown that premature babies did better under certain LED lighting conditions and that lighting could be useful in treating mood swings related to seasonal changes in sunlight. But there were also counterhypotheses about the negative impacts of light on health, related to disturbed sleeping patterns and the need for dark skies to reduce light pollution in cities. The industry had been conducting research since 2000 about the effects of lighting on human health, but two opposing views were now developing about the impact of white light from LEDs on human health. Only recently has light been viewed as an integral driver of circadian rhythm in humans. Also, its potential impact on brain health, including such

afflictions as Alzheimer's disease, is getting more attention. Another key research area is the impact of lighting on plant growth. The ability to vary the color and brightness of LEDs is one way to control pathogens, disease, and rot in produce, as well as to foster better growth and plant health.[7]

Philosopher Charles Peirce argued that neither *inductive* logic (reasoning from specific examples to general principles) nor *deductive* logic (from general precepts to specific truths) can truly generate new insight.[8] Instead, he proposed *abductive* reasoning to allow for "logical leaps of the mind" that go beyond deductive necessity to find fresh new explanations.[9] Abduction seeks to create "new facts" by using existing ones plus creative inferences. This approach is especially important when facing ambiguity because there are often many possible explanations or conjectures for the incomplete issues.

The aim of abduction is to generate new knowledge by steering the generation of competing hypotheses, by abandoning old convictions and seeking better ones. As strategy expert and author Roger Martin has observed, "When you are facing something that doesn't obey the previous rules or have some data (but not enough to be inductive), you make an inference to the best explanation of what is going on."[10] This logical leap of the mind leads to a new idea, and that idea can only be proven to be right or valid by the unfolding of time and future events. Put simply, abductive reasoning encourages organizations to pursue ideas beyond the tried and true. This means at times stepping into the dark corners of ignorance while modulating reasonable doubts for positive effects.[11]

Facing resistance to the adoption of LEDs during the early years (circa 2003), a few business units within Philips Lighting used abduction to spur action. They made bold, scary statements, for instance, about LED technology replacing incandescent and halogen lighting. Furthermore, they helped spur this new trend through high-profile marketing moves. For example, the New Year's Times Square Ball in New York had been lit with incandescent lighting since its launch in 1907. Philips helped arrange for this famous ball to be lit by LEDs as it descended in the final countdown during the last minute of 2011.[12] This very notable LED application became the talk of the town and far beyond. Many players in the value chain paid attention. Philips launched other high-profile projects as well, such as lighting up city nightscapes and famous building facades to demonstrate the broad applicability, aesthetics, and reliability of LEDs.

Explore Divergent Views

Because LEDs were projected to affect the entire lighting industry, it made sense for leaders to join forces with other firms to explore the challenges more deeply. Still, most lighting manufacturers were slow to spot the early warning signs of this technology substitution due to being one step removed from the customer. They were just not paying enough attention to the periphery, where all the early action for LED lighting was happening. The core of the lighting business was about producing bright and warm white light. This was the last frontier conquered by LED, which started with colored lighting in niche markets because the chemistry and physics of solid-state semiconductors could more easily be applied there.

A traffic light requiring three colors was a natural application for LED. But traffic signals were a sideshow for the major manufacturers: the primary customers were government agencies, which procured their replacement light bulbs from midsize or large distributors. These wholesalers operated in a highly competitive market segment with low margins, so some started to offer low-cost replacement (LED) lamps from manufacturers in China. At first there was only a limited supply, but around 2003 LED became a standard option for traffic lights, given its lower cost. The longer life of LEDs saves labor for utility workers and police officers in the rare case of traffic lights failing. However, this particular LED application was just one of many possible niche markets, and because it didn't involve white light (just color), the majors noticed it too late and lost this business. Circa 2005, Philips, Osram, and GE became more interested in general street lighting because it concerned bright white light, which was closer to their core business. By the time the majors fully recognized the cost savings, better visibility, and health benefits of LED,[13] the streetlight conversion to LED was well on its way. On top of this, there were smaller OEMs such as Dialight and Cree that launched integrated fixtures for system replacements, hurting the traditional lamp manufacturers further.

This example is a classic case in which the multistep channel process dampened the early warning signs, and the majors were unable to act fast once these signals became stronger. In the end, the majors were almost entirely eliminated from the traffic light segment and collectively lost nearly $1 billion in revenue in the United States alone, while incurring significant losses in street lighting as well. Similar stories played out in such application segments as exit signs, outdoor video displays, architectural lighting, backlights for flat panel (television and computer) displays, flashlights and refrigerator lights, and eventually indoor and outdoor white lights broadly.[14] LED at first conquered niche markets, often in a zigzag manner, making it hard for the majors to get a fix on its future trajectory. As the threats—as well as opportunities—of LED became increasingly clear, however, the incumbent players teamed up to create an innovation initiative called Bridges in Light.

Starting in 2003, the major stakeholders came together to map out the future of the industry and launch a "burning platform" to drive industry change. They developed wide-ranging scenarios for the future of lighting, some scary and others more promising. The scary scenario was intended to shake the industry out of its slumber, which would be cheaper and faster than waiting for a big crisis to hit. The scenarios described ahead were originally created as part of a strategic planning process within Philips, with one of us (Paul) guiding the effort. These in-house scenarios were the starting point for the Bridges in Light initiative at the Lighting Research Center (LRC), which adopted an industry-wide perspective.

The resulting scenarios were later used by a major industry trade group, the National Electrical Manufacturers Association (NEMA), to launch an industry-wide education effort, along with a broad marketing program to help the industry overall adapt to LED. One positive result was that the industry moved rapidly to include the emerging LED technology in its standards setting. The US Department of Energy (DOE) also joined the fray, aggressively advocating for further market transformation and specifically challenging the industry to develop high-efficiency LED light bulbs. For example, the DOE launched the $10 million L Prize to replace two of the most commonly used light bulbs at the time, the 60 W bulb and the PAR 38 bulb. These interactions with other industry stakeholders deepened the understanding of the impact of LED going forward.

While most of the lighting industry remained on the sidelines about LED, senior Philips executives developed mechanisms for internal discussion. A new Solid-State Lighting business unit was created to develop the LED market and devise appropriate strategies, with P&L (profit and loss) accountability. Because LED called for new capabilities, Philips entered into joint ventures, such as buying Agilent Technologies' stake in Lumileds in 1999 for nearly $100 million, to design and manufacture LEDs.[15] These and other technology investments benefited from Philips' well-known brand and broad channel access to markets. Its new LED division allowed Philips to forge ahead of the industry, although the internal process of debate to resolve conflicting priorities was often laborious and bureaucratic. Philips' corporate board needed to step in at times to ensure constructive debate, strategizing, and execution. The new internal LED group was fighting an uphill battle against powerful traditional business units in Philips that still favored investments in the conventional technology.

Develop Multiple Scenarios

Armed with the outcomes of the first three tools, scenario planning enables better understanding of the uncertainties that remain.[16] A useful set of scenarios offers diverse narratives about what the future might bring, organized around the main uncertainties. Each scenario narrative must be coherent, compelling, and insightful, with at least some deep challenge to the prevailing organizational mindset.

Each individual scenario should present just one internally consistent story about a relevant future that *might* emerge. By considering multiple such scenarios side by side, the organization is prevented from locking into only one view of what the future might bring while sharing a common set of frameworks for interpreting ambiguous signals.[17] Whereas organizations normally filter *out* weak signals from the periphery—especially those that don't fit the dominant worldview—scenario planning *amplifies* weak signals. In combination, these signals can then help foreshadow fundamental shifts in the marketplace and society at large. Because there are *multiple* scenarios in which a particular weak signal may have varying degrees of strategic significance, the organization avoids the trap of pushing important strategic uncertainties under the rug. Developing scenarios greatly helped Philips and others break out of narrow views of the lighting industry.

Because LED technology continued to make inroads into traditional lighting segments and beyond, incumbent firms realized they had to monitor the entire value chain to fully see all threats and opportunities. The traditional industry view, as depicted in figure 5.2, suffered from serious myopia and blinders. It focused on the end consumers, who mostly cared about the price of the light bulbs, resulting in a low-cost competition. The lighting companies also tried to influence the OEMs, specifiers or designers, and contractors who traditionally made the lighting choices for end consumers. With brutal price wars afoot, as well as battles for shelf space and mind share among contractors and designers, there was plenty to keep senior leaders occupied. The narrow industry scope shown in figure 5.2 obscured important parts of the bigger picture and fueled a race to the bottom.

A much broader lighting ecosystem view, such as that in figure 5.3, would have helped to highlight factors that could accelerate the emergence of LED technology or perhaps even slow its progress. For example, the spread of LEDs depended on technology, changes in customers' buying behavior and needs, and new regulations. The development of this new technology could also be affected by a variety of other forces, including health, security, technology, energy dynamics,

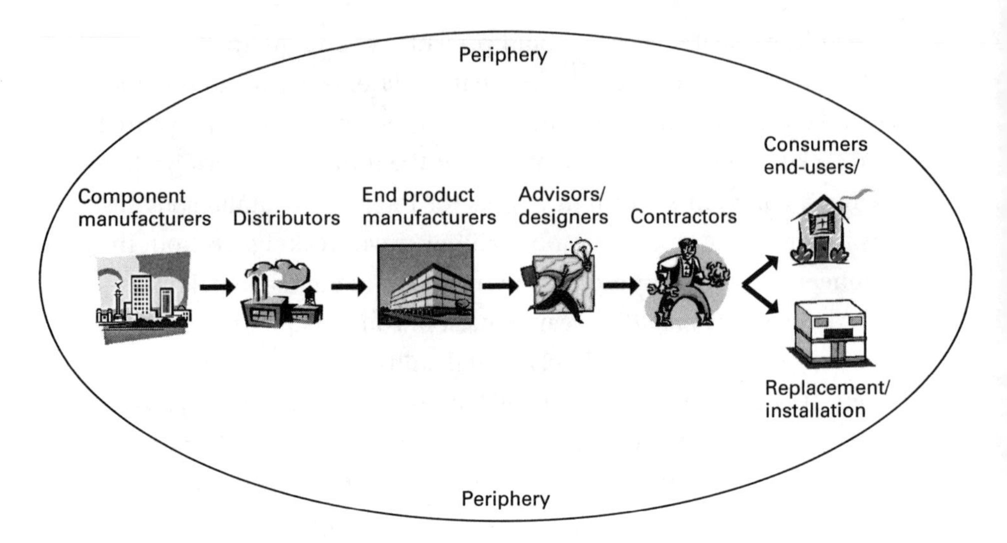

Figure 5.2
Narrow view of lighting industry

transportation, coordinated research, aesthetics, sustainable development, and dark-sky-related environmental issues.

How Philips Tackled Ambiguity

Between 2000 and 2005, Philips embraced each of the tools described in the previous section in an effort to gain firsthand experience with emerging solid-state technologies, from LED candles to ambient lighting systems for hospitals. "We used a launch-and-learn strategy to better understand solid-state lighting, as well as try out new business models," said Govi Rao, who was vice president at Philips Lighting at the time.[18] "These forays allowed us to monitor many factors such as channel conflicts and cannibalization effects. This is where incumbent companies are often blindsided. By creating pilots, we minimize risks. And if we make mistakes, we keep them small and learn quickly."

In 2005, Philips was especially interested in two key uncertainties. The first concerned the shift from conventional lighting—in which the company was strong—to truly new applications. LED would likely enable genuine lighting innovations due to LEDs' small size, long life, low

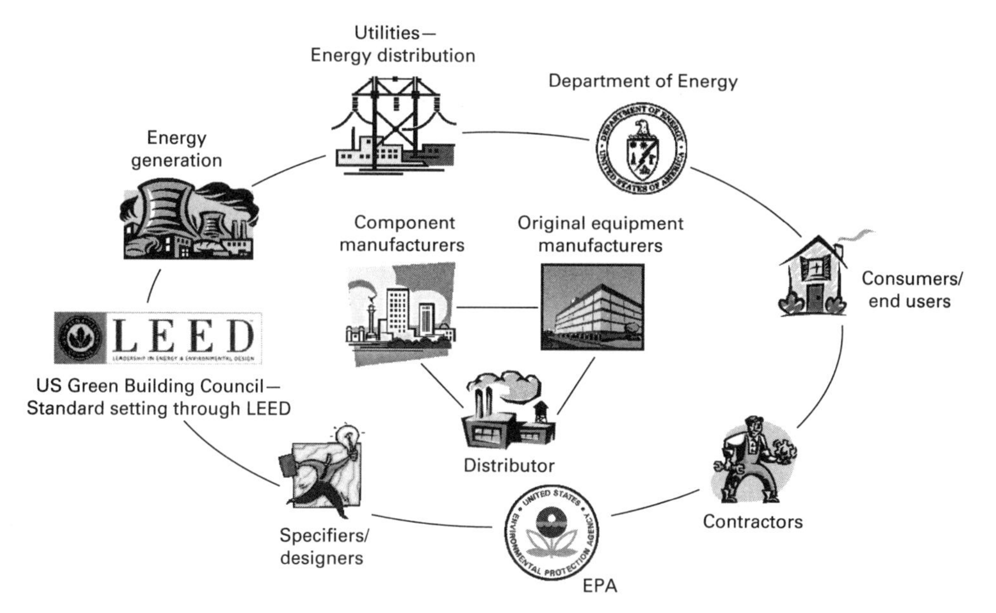

Figure 5.3
The broader lighting ecosystem

voltage, and wide temperature range, allowing them to be embedded in all kinds of materials. For example, a sweater might have LEDs embedded in its fabric and flash a smile whenever the wearer says, "Nice to meet you." More serious applications could see LEDs lighting urban landmarks, enhancing medical applications, and enabling military technologies.

Beyond novel uses of lighting, Philips was also interested in how these lights might be controlled by users (the vertical axis in figure 5.4). The era of physical switches affixed to a wall in either the on or off position was coming to an end because LEDs can be controlled digitally and remotely. Traditional lighting already offered dimming options or a clap of human hands to switch on a light, but nothing as versatile as the nearly infinite possibilities for LEDs. A challenge remained, however, in keeping product control options manageable for users at home and at work, without complicated instructions. It was still unclear how much digital sophistication consumers really wanted to pay for.

These two key uncertainties painted four starkly different narratives about LED technology's future for the period 2005 through 2015, as

		What are some new LED lighting solutions?	
		Modest (<10% market share)	**Extensive** (at least 30%)
What is the role of digital lighting controls?	**Limited** (mostly still hardware controls)	**Replacement with LED**	**Embedded solutions**
	Pervasive (digital remote controls win)	**Designers' delight**	**System-wide intelligence**

Figure 5.4
Scenarios centered on LED technology's impact

summarized in figure 5.4. These ten years could bring little change in either uncertainty dimension—corresponding to the top-left scenario, in which LED lights mostly replaced incandescent bulbs or fluorescent tubes. Consumers would still be intrigued by changing the light color or intensity via digital controls, but in this scenario that would remain a sideshow. The top-left scenario would be the least threatening for players like GE, Philips, or Osram Sylvania in the United States.

A more serious challenge would have arisen if LED applications significantly expanded the market with new, embedded solutions that would be out of reach for various incumbents (the top-right scenario). Controls hardware (as opposed to light sources) had always been a high-margin business but would undergo rapid change in this cell. Controls were mostly a components business, with few firms offering complete solutions. The top-right scenario would thus amount to a serious loss of profit due to cheaper LED replacements for existing lighting plus the loss of growth opportunities in new market segments. More challenging yet would be the bottom-left scenario, in which hardware controls for lighting (e.g., fixed switches on walls) give way to digital controls operated remotely from anywhere at any time.

Lighting controls were an important part of the market in 2005, with many competitors vying for a piece of the action. The bottom-left scenario depicted a *designers' delight* world due to new software applications and features far exceeding traditional hardware controls. Architects, designers, and interior decorators would gain in power once new, versatile controls reached attractive price points for consumers. The bottom-right scenario would be the most challenging future for incumbents, due to a switch towards intelligent, integrated lighting solutions that use very little of the old system.

To better understand the four scenarios in figure 5.4, Philips undertook various sensing and seizing activities. First, it launched a better environmental scanning system to monitor newly emerging market segments, as well as levels of LED penetration in the industry's traditional segments. Second, a task force was formed to review Philips' current patent position in LED technology and explore ways to broaden these patents in terms of applications and stronger legal protections.[19] Concurrently, the patent portfolios of key rivals were examined, as well as those of new entrants to assess threats and explore new partnering opportunities. Third, special senior attention was devoted to China's growing role in the worldwide supply of lighting products, as well as various moves being made by new Asian players with relevant semiconductor know-how.

Lastly, a major effort was made to spot early signs of new functionalities being launched, especially those protected by strong patents or complex software platforms from which further intelligent solutions could emanate. For example, Philips decided to strengthen its rather weak position in the U.S. market for armatures and focus more on the professional market for luminaires generally since both were viewed promising segments for LED penetration.[20] Using these insights, the company pursued a multipronged approach to lead the market: pushing organic growth targets, developing an aggressive M&A strategy, and building a portfolio of small bets that could eventually result in a couple of big bets. This combined approach allowed Philips to improve its offerings to the trade without directly competing with the company's very important OEM customer base.

As noted earlier, Philips also designed various experiments, such as ambient lighting for a new pediatric hospital pavilion that integrated various lighting technologies to deliver an improved CT scanning experience for the kids and their parents. Young patients could choose one of four different themes: aquatic, space, fly-through, or a default lava lamp. This helped keep the children still during a scanning procedure. If a child had to hold his or her breath for a procedure, an otter in an aquatic scene might do the same to set an engaging example. "We were learning by doing and creating strategic options based on what we'd learned," Rao said. "The value of such experiments lay in our ability to challenge the current paradigm of doing business." These and other moves positioned Philips to weather various threatening storms on the horizon while spurring innovation. By 2017, LED technology accounted for 24 percent of Philip's lighting sales and placed it in the lead position compared to its traditional rivals.

In retrospect, the onslaught of LED was not as massive as envisioned in the bottom-right scenario, but it was still a very significant game changer. Depending on the market segments and the local conditions in different countries, the industry was operating in either the top-right or bottom-left scenario in 2015 (ten years out from the 2005 scenario projections). Even though the scenario matrix in figure 5.4 is now fifteen years old, it's still conceptually relevant because lighting is expanding beyond providing better illumination. Horticulture is an especially promising growth area. When plants are grown in vertical farms, where light can be precisely regulated, the crop yield and uniformity of fruits or vegetables increases significantly. Targeted lighting has become a nonchemical method for controlling plant disease and a way to increase the nutritional content of leafy crops.[21]

The Philips example illustrates how companies can use scenario planning to deepen their understanding of new technologies or markets while not just betting the farm on a single future narrative. In the end, Philips decided to exit the low-margin lighting business altogether. Three of its scenarios pointed to serious value migration away from its own business model. With LEDs embedded in myriad devices, with cheaper building materials and casings, the margins became too thin for traditional light bulb manufacturers. By moving ahead of GE and Osram, Philips still had time for an orderly IPO in 2016, followed by the sale of its remaining lighting business for several billions of euros.[22] The margin squeeze that Philips saw in time resulted in the breakup of a long-standing dominant oligarchy. All three top players are now exiting the lighting business.[23]

Conclusions

Interpreting ambiguous situations requires a high degree of doubt about the current problem framing and proposed solutions. Unfortunately, leaders who express doubt may be viewed as indecisive, and doubt can indeed be dysfunctional if it leads to analysis paralysis. *Constructive doubt* actually helps to instigate inquiry and resolve ambiguity in support of better decisions. As biologist Stuart Firestein has emphasized, the domain of ignorance is far larger than that of knowledge and therefore far more interesting for garnering new insights, whether in science or business.[24]

Leaders can nurture a vigilant mindset by fostering a climate in which doubt is acceptable. One way to sow constructive doubt is to encourage colleagues to voice their hunches, which may manifest themselves in the form of fear, irritation, and boredom, or perhaps even feelings of guilt. As one expert phrased it: "When we nurture hunches, we cultivate the generative potential of doubt."[25] Once doubt exists, the crucial process of inquiry and testing can begin, motivated by boundless curiosity about new questions and challenges.

Another approach to fostering productive doubt is to explore surprises—positive and negative. If lessons are extracted, every mistake has a silver lining; likewise, every unexpected success can offer a further bonus via deeper insights. Once many iterations of these learning loops have been completed, team members will start to view ambiguity and doubt as critical to strategy formulation and deserving of respect and trust. Once ambiguity has been confronted and sufficiently resolved, the time is ripe for making the strategic moves discussed in the next chapter.

6 Taking Timely Action

Quick decisions are not the safest.

—Sophocles[1]

Leaders considering potentially risky moves in turbulent environments may hit dead ends because of incomplete or biased information. Vigilant organizations recognize that it seldom pays to go all in with new strategic initiatives. But they also know that this harsh reality can be tempered with a robust sequence of actions. First, they can convert signs of possible risks or opportunities into hypotheses to be tested with an experiment. We're not talking about carefully controlled scientific experiments, but rapid and iterative tests that build on the learning process explained in the previous chapter. Second, they can invest in strategic options that serve as smaller bets and can be quickly unwound if things don't work out. Third, they can extend their capabilities and contain risk further by working with a partner facing a similar situation.

These three actions can liberate the firm's leadership from the false dichotomy of either overcommitting in haste *or* doing nothing while waiting to see how things play out. Rather than embracing an all-or-nothing approach, these staged commitments allow a leadership team to move onto new opportunities sooner and avoid defensive overreactions later. A flexible approach to taking timely action is especially well suited to established players facing possible digital disruption—such as companies in the hearing aids and life insurance industries. Let's take a closer look at the challenges faced by each.

Hearing aids. Six major American suppliers, including Oticon and ReSound, have long been protected by regulatory barriers such that only trained audiologists can evaluate, install, and service hearing aids. These devices cost from $900 to $4,000 per ear, depending on the extent of hearing loss. Because insurance seldom fully covers the cost, only one in five people who could benefit from a hearing aid actually wears one. On the horizon are two significant threats to incumbents.[2] First, the US Congress is considering a deregulation bill that would permit over-the-counter sales. This would open the market to numerous technology companies, including Samsung with its Gear IconX earbuds, Apple with its AirPods, and Bose with its Hearphones that were selling for $499 per set. Second, in an effort to avoid FDA oversight, these players—plus dozens of new companies piling into the expanding space—are emphatic that their devices are not actually hearing aids. With prices up to 90 percent lower than traditional aids, the market will surely expand when they enter it. How should the incumbents respond to these threats to their stable and highly profitable business?

Life insurance. Few industries are as staid and seemingly impervious to digital threats, but digital technologies can greatly improve customer value, make policies more relevant, or remove the costs of intermediaries.[3] Traditional firms—relying on agents with hefty, up-front commissions that can be more than 100 percent of the first year's premium—can be attacked by entrants using automated advisory processes and advanced analytics to sell term insurance directly to consumers. These start-ups could use second-party data services such as prescription histories to issue policies without requiring customers to undergo a time-consuming medical test, and some offer big discounts to policyholders leading a healthy lifestyle. It is only a matter of time before outsiders like Apple will leverage their data and product platforms to enter the market.

Most firms today face some variant of these same challenges, and playing defense is not an option. The techniques for interpreting ambiguous issues discussed in the previous chapter lay the groundwork for a three-pronged approach to *acting faster*.

Step 1: Run Trial and Error Experiments

By 2011, the traditional "share of voice" model of selling pharmaceutical drugs—which was all about how much time a salesforce could spend with doctors compared to rivals—was crumbling. The digital moves made by Novartis Pharmaceuticals in the next three years show the gains possible from preemptive moves guided by trial and error experiments and deep customer dialogues.

Historically, pharma sales representatives made visit after visit to prescribing physicians, following a carefully constructed script. These interactions were brief, one-way communications that often felt—to both the sales reps and clients—like monologues. As a result, Novartis had little data on how its drugs were being perceived and used by its customers, and few ways of knowing which sales strategies were most effective. Throughout the "fat years" of blockbuster drug breakthroughs, handsome margins meant there was no need for pharma companies to reinvent their standard approach. But eventually, most drugs lost their patent protection and generic versions began to capture market share, leading to sharp price cuts. Simultaneously, purchasing power and influence shifted to payers or providers, while the pharma companies were criticized for dubious sales practices.[4] Sales representatives found themselves in a radically changed world: their access to physicians was increasingly restricted, and the number of new products that they could discuss—if they managed to get in the door—was dwindling.

To respond to declining sales and the mounting salesforce frustration, Novartis launched a digital initiative in 2012 to help its twenty-five thousand sales representatives in eighty countries engage with doctors in consultative two-way dialogues.[5] Value-added services and broad channels of communication replaced the simple recitation of standard messages. By involving the most relevant scientific staff, answers to complex questions could be provided in real time. Reps could immediately access whatever data the doctor would find most useful, whether it was safety issues for the elderly or the risks of drug interactions. This allowed conversations to flow more naturally in two directions, putting an end to formulaic sales scripts.

Moving from monologues to dialogues allowed Novartis to sense weak signals from these new conversations and improved its understanding of the factors that mattered most to customers. Its digital platform captured detailed information about sales visits, improved understanding of customer preferences, and enabled personalized marketing messages. Novartis was able to spot budding problems early and sense market opportunities sooner. The spirit of trial and error learning further suffused the company's customer-engagement initiatives. As many as forty-two pilot tests were conducted in multiple countries to learn how best to design the platform, decide which features to include, and monitor acceptance by sales representatives. The feedback from these experiments was crucial to the final investment-commitment decision. Organizational learning requires creating experiences in the real world and then using the results to scale up or down subsequent investments. Embedding this practice will elevate it to a dynamic capability that can be used in many ways—provided two conditions are met. First, the organization must nurture an experimental mindset, including a willingness to challenge existing beliefs. Teams employing experimentation must be able to codify and share their insights. New software tools, including advances in data analytics, can help teams keep track of test and control groups, as well as help identify the attributes that most affect performance. Second, leaders must nurture a culture in which mistakes are tolerated and even judiciously encouraged. Although careless or negligent failures should always be avoided, no organization can learn if it pursues a policy of zero tolerance for failure. It isn't enough to pay lip service to the idea that mistakes will be tolerated; it takes deliberate effort to foster a climate that learns from failures and in which experimentation is built into the mindset. As Einstein noted: "If you have never failed, you never tried anything new."

A firm with a robust tolerance for failure has leaders who recognize that every setback offers the potential for insight. For example, leaders can conduct postmortems for major projects and investments that didn't succeed, aware that the greatest insights often lie on the far side of failure. Leaders should not just praise success, but also recognize and laud projects that failed—boldly, perhaps, but for the right reasons. Danish company Grundfoss, the world's leading pump manufacturer, embraces this principle by asking assembly-line workers to document every time something *almost* went wrong. When these methods are practiced throughout a firm, it becomes a genuinely inquisitive enterprise.

Step 2: Invest in Strategic Options

Green technologies—defined as science-based applications that aim to conserve the natural environment and resources by minimizing waste and toxicity—are fraught with uncertainty. There are multiple emerging technologies involving significant capital risks to be managed over extended time periods, often via partnerships and strategic alliances.[6] When DuPont first sensed an opportunity to apply its biotechnology expertise to green technologies, it saw that many governments were being pressed to deal with climate change and energy-security issues.[7] It also realized that ethanol—an alternative energy source—was vulnerable to displacement because it cost twice as much to produce as gasoline and had a significantly lower energy content per gallon. To contain the risks of exploring alternative biofuels, DuPont deployed several strategic options, followed by a series of joint venture partnerships.

This approach, which played out over decades, illustrates the power of investing in strategic options.[8] Akin to financial options, these strategic options[9] preserve the right to invest more later without an obligation to do so, allowing for midstream adjustments once more has been learned. By developing a valuable portfolio of strategic options across many investments, leaders can then experiment, prototype, and refine their planning, depending on where markets and technologies are heading. Because this direction may not have been clear at the outset, the options approach lets companies win across multiple scenarios that *might* evolve.[10]

There are four types of strategic options (see figure 6.1), and most firms will need to assemble a portfolio of them:

1. *Preserve and protect options.* These are suitable when the market and technology spaces are familiar and uncertainty is manageable. They allow a firm to respond to possible competitive moves, shifts in market requirements, or surprises in the economic climate. Such options are created through carefully devised experiments that test different hypotheses or strategic responses and through anticipatory development programs that ensure the firm isn't left behind when rivals move.
2. *Disposable options.*[11] These options are used when a firm wants to pursue an opportunity quickly without building a large-scale production facility. They make sense when demand is uncertain and a large-scale plant might become an albatross if demand proves weak. By starting with a smaller plant (either alone or in partnership with others), there's an option to build a bigger plant once demand proves strong while avoiding long delays. If demand fails to materialize, any loss remains manageable, especially if shared with a business partner. In managing disposable options, it's important to understand the salvage value of any investments because disposable options face a greater risk of failure than other kinds of investments firms make.
3. *Exploratory options.* These are used to deal with high market and technical uncertainty until sufficient commercial feasibility has been established. These small, exploratory investments help a firm acquire additional experience that can later be parlayed into larger strategic commitments. Small R&D investments, joint ventures, or investments in start-ups serve this purpose. The acquisition of a small start-up, for instance, is an opportunity to test the waters, and it can be abandoned easily if it doesn't deliver the desired value. But if it succeeds, it provides a knowledge base and a framework for further investments.
4. *Scouting options.* These are cautious investments made to discover new technologies or markets when uncertainty is quite high. The military scouting metaphor is apt: to find an enemy, the army sends out scouts—and even if they fail to return due to capture or death, the generals will at least develop rough knowledge of the enemy's

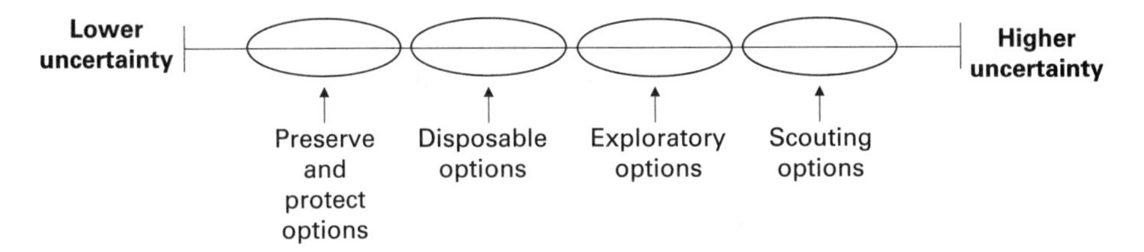

Figure 6.1
The spectrum of strategic options

whereabouts. Scouting forays can run the gamut from making small investments in start-ups to gain an early window into a new technology, to collaborating with think tanks or academic institutions to learn more about new technologies, markets, or tools. Another approach is to join benchmark groups, industry syndicates, or trade associations, provided they yield useful intelligence about the market, regulations, or rivals.

Step 3: Partner for Joint Gain

A do-it-yourself approach is often antithetical to taking timely action because few firms can overcome the constraints of inexperience, talent shortages, or financial limits without incurring lengthy time delays and risking too much. Novartis, for example, possessed deep capabilities in pharmaceutical drugs. To meet its strategic commitment to a digitally enabled salesforce, the company did not try to master the myriad digital disciplines it needed. Instead, it used development partners to design an innovative, nonlinear navigation approach that allowed sales reps to traverse through different content domains without ever losing eye contact with the physician. The company also collaborated with a cloud-based provider for its software, while graphics agencies developed the information architecture, file formats, and visual design.

Innovation and R&D processes were among the first areas in which firms could open up to external inputs and contributions. This shift was encouraged by the success of Procter and Gamble's Connect+Develop approach to developing new products.[12] P&G leadership recognized that "not all the smart people work for us"—that for every P&G researcher, there were probably two hundred scientists or engineers outside the firm who were just as good. It also realized that many of its best ideas in past years had come from teams working across divisional boundaries.

Being an open organization is not simply a matter of bolting on a few joint venture partners to enter new markets or outsourcing non-core activities. Rather, it demands a shift in mindset and strategy to give up some control and ownership to an external partner—putting a premium on relational capabilities that can extend the firm's resources beyond its own boundaries to access the complementary resources of partners.[13] The ability to manage multiple complex partnerships entails a complex set of strategic and tactical coopetition skills.

Between relying primarily on internal resources and opening up widely to mobilize a network of supporting partners, there is a broad spectrum of partner-to-gain options. Moving to the right on the spectrum in figure 6.2 requires cultural shifts toward collaboration and a willingness to share information, costs, and benefits. These are necessary conditions for moving speedily to participate in groups of firms developing digital systems for everything from smart homes and workplaces to real-time, multichannel financial transactions. Embracing this takes a mental shift from a firm-centric view of strategy toward a network or ecosystem perspective that is broader, more open, and fluid. There are three main ways to gain from partnering.

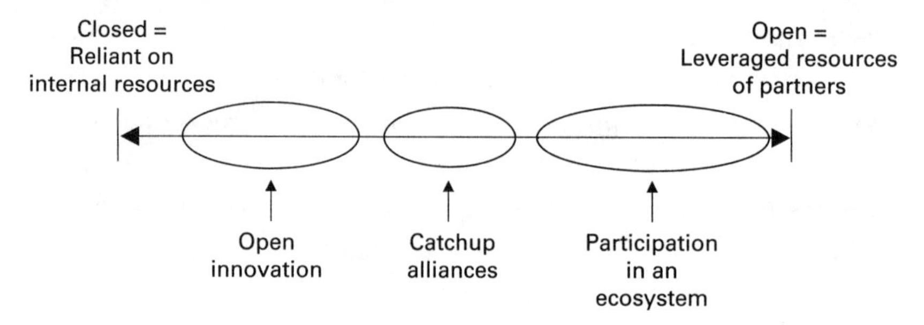

Figure 6.2
A continuum of partnering possibilities

Open innovation partnerships.[14] The move to open innovation is about searching outside the boundaries of the firm for new ideas. This approach was greatly accelerated by digital technologies for coordination of activities such as 3-D printing and shareable, quickly modifiable designs.[15] There are two facets to openness. The first is the outside-in aspect, in which external ideas, technologies, and capabilities are brought into the firm's innovation/R&D processes. The second is an inside-out aspect, whereby overlooked, forgotten, or underutilized ideas or technologies are identified, shared, and incorporated into others' innovation activities. Examples include L'Oréal and Renault working on an "electric spa" concept car, or auto suppliers Delphi and Mobileye collaborating to design an autonomous driving system. The interactions between firms are facilitated by idea-sharing systems such as Slack.

Catch-up alliances. These can range from consortia with flat governance approaches and closed membership that share technologies to long-term cooperation agreements made between bitter rivals to catch up to industry leaders. In these alliances, technology laggards[16] are forced to turn to each other and open up their innovation processes to survive. When these firms don't see and respond to opportunities soon enough—for example, joining the hybrid-electric drivetrain technologies consortium for automobiles, now led by Toyota—the costs and risks of catching up on their own become prohibitive. This type of alliance is risky because as they learn more, participating firms usually find their interests and motivations diverging.

Ecosystem participation.[17] Business models embedded within ecosystems are reshaping many traditional industries—and creating entirely new ones. To be successful, any ecosystem must offer compelling ben efits to potential partners and be applicable to multiple businesses. Vigilant firms can explore the possibilities of an emerging ecosystem by starting with a clear value proposition and broad perspective. For example, potential members should have clear (but not fixed) positions within the ecosystem, activity will flow between them, and they will be aligned through compatible incentives and motives.[18] Taking such an outside-in view of the various benefits that the ecosystem can potentially deliver—rather than focusing on what each partner can contribute—makes the coordination of joint activities both valuable and feasible.

The automobile ecosystem coming into view is a good digital example. It involves diverse players providing solutions centered on four technology platforms: autonomous driving systems; connectivity through telematics, voice recognition, and related technologies; electrification and energy storage; and shared mobility. The players include traditional first-tier OEMs and their suppliers, competing frenemies (parties sharing common enemies), and unexpected new entrants. This evolving ecosystem, like many others, will be formed as a thicket of interlocking alliances among strange bedfellows, few of whom are monogamous. For example, graphics chipmaker Nvidia is building embeddable computers for eight different producers of self-driving cars. In this fast-moving environment, vigilant firms are the best equipped to anticipate the new capabilities needed to bring partners together.

Combine Approaches Flexibly

The partnering possibilities in figure 6.2 are mutually reinforcing. When facing digital turbulence, there may not be enough time for trial and error learning, so it pays to leverage the experience of others. Apple did this through collaborations with Napster and other file-sharing services, as well as its own acquisitions, quickly gaining the knowledge needed to launch the iPod. Even so, Steve Jobs and Apple still nearly missed this mega digital disruption in music. In the summer of 2000, Jobs was still highly focused on perfecting the Macintosh's capabilities for video editing and almost didn't see one of the biggest revolutions in the music world. "I felt like a dope," he said later in an interview with *Fortune*. "I thought we had missed it. We had to work hard to catch up."[19]

Once Jobs recognized the enormous potential of digital music, Apple acted quickly by adding compact disk burners to all its computers. Then it purchased a small company, SoundStep, to jumpstart software development. The first version of iTunes was created in just four months, and the first iPod arrived nine months later. The company still needed content to put on its iPods, so it worked out critical agreements with major record companies to develop a platform for selling the songs. Fortunately for Jobs, the music industry was so busy suing Napster and its own customers—which slowed down its own digital innovations—that Jobs had more time than he realized. When Apple's iTunes music store opened in April 2003, the goal was to sell a million songs in six months. This target was surpassed in just six days, and by early 2005 iTunes controlled 62 percent of all legally downloaded digital music.

Keep it simple. The term *analysis paralysis* sums up the risk of adopting decision procedures that are overly measured, linear, burdensome, and slow. In stable situations, optimization approaches based on decision theory and operations research can work well, but not in the hotbed of Silicon Valley. The most successful companies there use just a few simple rules to stay ahead of the turbulence.[20] Effective dynamic capabilities in high-velocity markets are simple and not nearly as elaborate as they need to be in less turbulent settings.[21] Simple routines keep managers focused on what matters without locking them into specific behaviors.

One danger of overly analytical approaches is that they draw on past experiences that are misleading, akin to overfitting a statistical model to datasets containing old information.[22] When business becomes complicated, strategy should be simple.[23] The idea is to quickly move back and forth between hypothesis generation and testing in a series of rapid microcycles. Simple iterative probes are easy to apply quickly and, if guided by some wisdom within a narrow domain, they can do wonders.[24] This contradicts a common belief that mental shortcuts extract a steep price in the form of bias or suboptimality.[25] The ongoing debates about the wisdom of simple rules hinge on the kind of environments decision-makers operate in, as well as their cognitive abilities.[26]

Much Depends on Timing

In an era of digital turbulence, the right thing at the wrong time is the wrong thing. This cautionary statement resonates with firms seeking to find the right balance between cautiously learning about a digital opportunity and investing in a few strategic options on the one hand and making a major commitment to a digital transformation on the other.[27] The difficulties faced by LEGO (which has defunded its Digital Designer virtual-building program) and Nike (which ceased production of its FuelBand activity tracker) are cautionary. Meanwhile, the efforts of General Electric to become a digital industrial company by embedding sensors into many of its products and building an ambitious software platform for the Internet of Things, have been sidelined. In these and numerous other cases, the digital transformation ran into performance issues, causing the stock price to languish because big bets didn't pay off as quickly as expected. One lesson to be drawn is that instead of ramping up quickly (only to later ramp down painfully), it's better to make cautious moves and be ready when the timing is right.

Vigilant companies are confronting this challenge, as we discussed in chapter 3 when noting that the leaders of Shell were forced to refocus their attention on renewables. First, they recognized that multiple digital disruptions were afoot in both the *supply* of energy, such as batteries, and the *demand* for more efficient devices to conserve energy at home, at work, and while traveling. Clean electrons, for instance, might replace the complex hydrocarbon molecules that Shell knows how to find, surface, crack, distill, and transport around the world, at considerable environmental costs. Second, the new opportunities presented by clean energy do not fit the DNA of oil companies. Their hard-earned expertise in managing large-scale projects and complex technologies, dealing with governments, investing for the long term, and transporting energy may remain an asset. But the energy landscape is also rapidly changing strategically "from complicated to complex," according to Shell's executive for strategy, meaning that many of its core competences may become irrelevant.

Timing the cycles. The biggest challenges for Shell's leaders will center on timing. The company needs to monitor closely which way the winds might be blowing and then act fast enough to create meaningful strategic options for the future. To use the metaphor of a former Shell CEO, the key question is how many different pots to keep on the fire—recognizing that Shell reached into several technology pots for renewable energy in the past, only to get scalded. With solar panels, for example, it discovered too late that it was a cutthroat manufacturing game. Similarly, wind farms yielded margins deemed to be too low. The question now is how sharply the company's leaders should redirect their organization's attention toward clean energy.

In the past, Shell failed to give renewables enough strategic leadership attention, resulting in timid moves and sloppy execution. This time, the game change is for real: at least half of the energy system in the second half of this century will likely consist of renewables, calling for different competencies and navigating new value chains. In anticipation of this, CEO Ben van Beurden started to slash costs by redesigning deep water oil platforms and onshore shale-gas projects to succeed at lower prices. In addition, van Beurden has been investing in wind farms and joining solar consortia. The end objective is to push renewables through Shell's global network for producing, trading, and selling energy. Because the energy market is changing faster than Shell could have imagined, its leaders have to move quickly to meet this enormous task. They are again betting on hydrogen and joined a German experiment to install over four hundred hydrogen-fueling stations across that country. Shell also entered the battery-powered car industry by buying the Dutch company NewMotion, which operates thirty thousand electric-car-charging stations in Europe.

Shell's CEO is keenly aware of the challenges ahead; his famed scenario-planning team clearly laid out a world of "lower forever"—which could mean sitting on massive, unprofitable hydrocarbon reserves.[28] In response, Shell sold most of its stake in Canadian oil sands, a business closer to strip mining than oil exploration. Based on its future energy scenarios, the company expects that global demand for oil might peak in as little as a decade—which is very near term for a company accustomed to making multidecade investments involving billions of dollars around the world. Other Shell scenarios project peak demand for oil to happen late in the 2040s, so much uncertainty remains around the timing. But the company realizes full well that an epic inflection point lies ahead that will transition the world from petroleum to clean electricity.

Hype cycles. Leaders of technology companies often use the Gartner Hype Cycle to subdue those who are overly enamored of seductive technologies emerging from the periphery.[29] Although this framework has not been rigorously tested academically, it's a useful reminder that technologies can have very long time horizons. However, this is neither a justification for contentment with the status quo nor a reason to jump the gun and overcommit to investments. The hype cycle can help fine-tune a timing decision because it describes a five-stage pattern that digital technologies generally follow (see figure 6.3):

1. A *technology trigger* introduces new possibilities that capture the collective imagination and create an early stage of excitement (such as "big data is the next big thing!").
2. This enthusiasm quickly reaches a *peak of inflated expectations* that overshoots the reality of what the present generation of technology can deliver (as with the dot-com boom and bust).
3. As reality sets in, expectations drop and the market slips markedly into a *trough of disillusionment* (as happened with gene therapy after setbacks and a moratorium on human testing).
4. Still, the technology continues to improve through experience effects on the processing side, creating a *slope of enlightenment.* For digital technologies, this happened thanks to massive data collection and advances in the cloud-based big data machine-learning infrastructure. This is the point at which expectations rise in alignment with what's actually achievable at a reasonable cost.
5. Finally, once reaching the *plateau of productivity* phase, the technology is absorbed into daily life (as with iPhones) and grows steadily—until the next disruption arises.[30]

The predictive power of the Hype Cycle framework—in terms of timing and magnitudes—remains open to question. But its underlying phased dynamics are sound because they combine two different but related phenomena: the *rate of adoption* and *enthusiasm for a new*

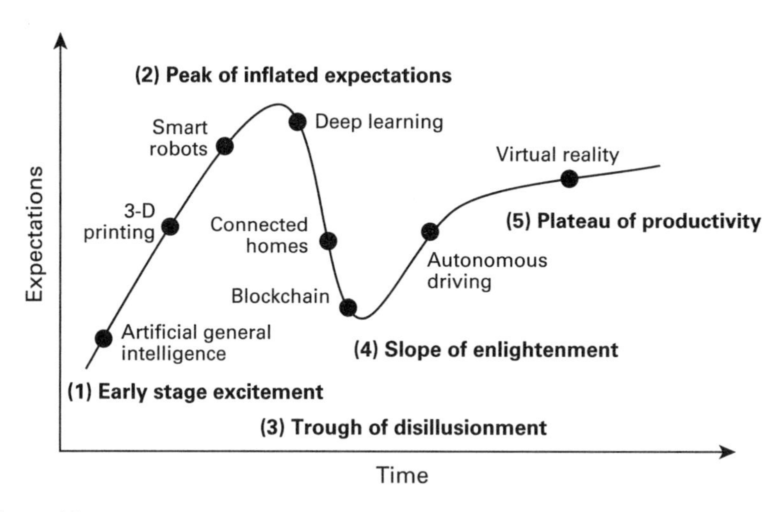

Figure 6.3
Hype Cycle dynamics and projections
Source: Gartner Group

technology.[31] Because either component is uncertain, these two dimensions provide a useful framework for scenario planning. Leaders should remain skeptical, however, of overly deterministic cycle predictions and think instead in terms of broad bands of uncertainty around each stylistic projection. The path of emerging technologies is notoriously hard to predict, with many twists and turns. The very label of *hype* suggests that irrational forces such as excessive hope, fear, or greed drive the market cycle, but it isn't always clear how much irrationality is at play. Akin to economic bubbles more generally, a hype cycle can correctly signal an emerging technology worth paying attention to (just as investing in the stock market prior to the dot-com bust in 2000 was initially a good idea). As the cycle concept emphasizes, the trajectory of an emerging technology is not quite a random walk but still embodies much uncertainty.[32]

The first relevant message is that the emergence of a technology in a hype cycle should not be ignored because *not all* of it will be hype. Hence, some senior-level attention should be allocated to the emergent digital technology to monitor its progress and strategic possibilities.

Second, the leadership team needs to develop a capability that is deep enough to appreciate what the technology really could mean commercially and be willing to probe and poke further. This means developing a long-range strategy for recruiting new talent and establishing a pool of in-house resources and outside experts. In areas like AI—where talent is scarce and the domain is diverse and deep—it may be better to build on a network of service providers who can help a firm navigate the uncertainties of the technology. In any case, ensuring that the right capabilities are in place is a precondition for successfully acting faster.

When dealing with turbulence, the root challenge remains balancing the competing urges to proceed very cautiously versus aggressively overcommitting. Leaders often get caught on one of the horns of this enduring dilemma and thus need above all to remain vigilant. In the next chapter, we present a practical agenda for integrated actions by leadership teams.

7 Vigilance: An Agenda for Action

> So give me a turbulent world as compared with a stable world and I'll want the turbulent world.
>
> —Andy Grove, CEO Intel Corp[1]

Few businesses are more exposed to the crosscurrents of digital turbulence than the credit card industry. Between technological advances and the unpredictable preferences of millennials, the landscape is changing on an almost weekly basis. Yet as with most industries, not all incumbents have been equally vigilant. To its credit, Mastercard was the first to recognize the challenges posed by new digital platforms and moved quickly. By seeing sooner and acting faster than its competitors, it has been able to position itself as the preferred partner for many erstwhile disrupters.[2] Mastercard's transformation is an instructive example of the power of vigilant leadership guided by a disciplined agenda for action. The particular features of its journey were shaped to the situation, but the underlying design principles apply to other firms seeking to keep their footing amid digital turbulence. They serve as a broad action agenda for champions of vigilance.

The purchasing process enabled by Mastercard might appear seamless when you order a latte at Starbucks, but it takes a myriad of digital activities behind the scenes to authorize, authenticate, and settle each transaction. This complex infrastructure for high-velocity data processing and security once seemed a formidable firewall for incumbents against outside attacks. But by 2010, there were early signals of danger: Google was exploring payments with Google Wallet; Amazon and Apple both had payment projects underway; M-Pesa, launched in 2007 by Vodafone in Kenya, had become the world's largest mobile-phone-based money transfer and financing service; and WeChat, founded in China, would soon startle the payments world by signing up one hundred million users for its digital wallet and payment service within *one* month after launch.

Prior to 2009, Mastercard had focused its efforts on strategy execution, meeting quarterly earnings targets and catching up to its archrival, Visa. When Ajay Banga became CEO that year, following thirteen years with Citibank, he found an accomplished but laid-back organization, without a strong sense of urgency. Rather than launching yet another "burning platform" of looming digital threats, Banga energized the organization by focusing on untapped opportunities. As he said later: "The activities of the PayPals, the Googles, the Visas, the Am Exs ... actually make up only 15% of the world's retail purchases. A whopping 85% of all global purchases (and in the U.S., a still sizeable 50%) are in cash or checks.... If all you do is focus on the 15%, you just put yourself in a box that's smaller than what you're capable of dealing with.... So we redefined our competitors. Once we got (that) right then the vision of the company became very clear."[3]

Job one for Ajay Banga at Mastercard was to inject urgency throughout the firm. He was shocked to hear people say: "Yeah, we could have done that deal ... Well, we didn't really want it. The pricing wasn't right." He quickly made it clear that "there is no such thing as a deal you do not want to win."[4] When describing his own approach, Banga highlights key aspects of being a vigilant leader:

Have the courage to take thoughtful risks. "Rarely are you going to have perfect information. The thoughtful part depends on your humility and realizing you don't have all the answers—that you can learn something from everybody."

Be competitively paranoid. "Constantly ask yourself if you're missing something ... you need a sense of self-introspection to be a real leader."

Develop a global view. "Increase your connectivity to the world around you ... the key is to go beyond looking at the world through the lens of your company, your organization or even your country."[5]

These entrepreneurial features of Banga's approach to leadership shaped how he systematically transformed Mastercard into a more vigilant organization. Although this had not been one of Mastercard's stated goals, his approach set the stage for the company to become an exemplar of vigilance.

Becoming More Vigilant

Most firms have the desire and the potential to become more vigilant but are unclear about how to make it happen. Should they adopt a freewheeling, opportunistic approach, or a tightly disciplined process? Which features of the most successful change initiatives observed elsewhere might apply to them? Where and how should the leadership team start the initiative? In the preceding chapters, we identified some common themes of effective change programs. Here we bring them together in the four-step change process depicted in figure 7.1, which serves as a general roadmap for answering these early design questions.

This guide is less orderly and more freewheeling than most change processes, but it always begins with visible demonstrations of leadership commitment. This is reinforced by three further stages that put the leadership team's intentions to work: alter how strategy is formulated, make investments in foresight, and finally align the organizational activities. These implementation steps may proceed in sequence or in parallel as the situation dictates, with variable pacing and priorities. What worked for Mastercard—or Adobe Inc., as we saw earlier in the book—has to be adapted and modified for each case. There are many iterations and feedback loops in any learning by doing, multiyear change process. But as Oscar Wilde once said about history, "It may not repeat itself, but it surely rhymes."

The change process shown in figure 7.1 combines lessons from our research into the drivers of vigilance described in chapter 2 with

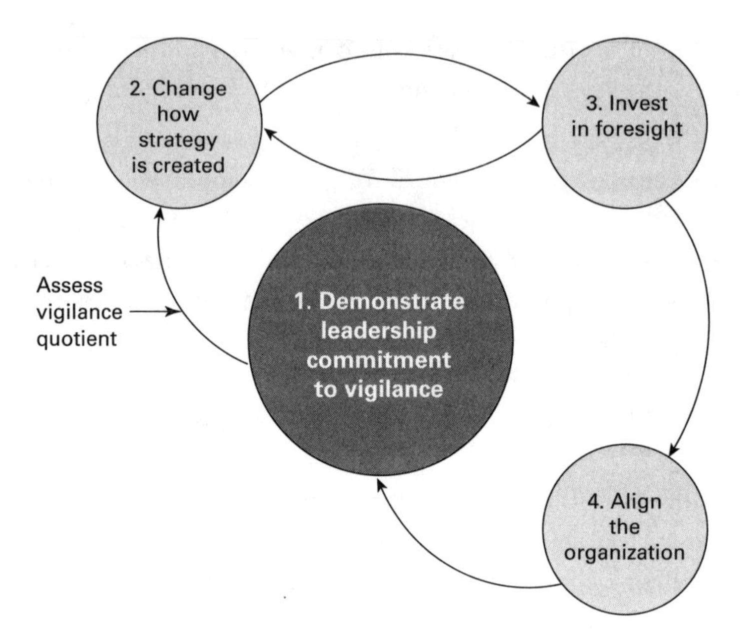

Figure 7.1
Becoming more vigilant: an action agenda

hands-on experience working closely with executive teams during the past decade.

Step 1: Demonstrate Leadership Commitment to Vigilance

The starting point is to have the CEO and the leadership team ponder six "leading" questions. The answers uncover whether the team is serious about becoming more vigilant or is just going through the motions. For each question, there are well-tested answers to help strengthen the vigilance capability and demonstrate commitment throughout the organization.

How Much Time Do We Spend Exploring the Future?

Short termism is the number one enemy of vigilance about longer-term undercurrents. This key point was reinforced by the findings of the CEO Genome project's assessment of two thousand CEOs.[6] Leaders who excel at adapting to turbulent environments spend significantly more time—as much as 50 percent of their time—thinking about the long term to spot subtle trends. Less successful chief executives devoted only 30 percent or less of their time to long-term thinking. The conclusion was that a long-term focus makes CEOs more likely to pick up earlier on weak signals and sense potential changes.

Do We Encourage Diverse Inputs?

For example, Ajay Banga championed diversity from the start, arguing: "Leadership attributes are tremendously facilitated if you surround yourself with people who don't look like you, don't talk like you, and don't have the same experiences as you. Diversity is essential because a group of similar people tend to think in similar ways, reach similar conclusions and have similar blind spots. To guard against that, you need to harness the collective uniqueness of those around you to widen your field of vision—to see things differently, to fail harder and to question everything. Widening that field of vision means widening your worldview."[7] Leaders can encourage vigilance by hiring for it as well as by promoting "mavericks" or others from within who are naturally more vigilant and externally focused. Mavericks tend to be independently minded.[8] They hang out with different groups and read and digest different media than their peers. When hiring new staff, specific questions can be posed to assess a person's ability to scan without taking his or her eyes off the main focus, and when designing performance reviews, leaders should evaluate how frequently and successfully employees have picked up weak signals from the periphery and reward them for it. Vigilant leaders should also educate employees in critical and divergent thinking skills, scenario planning to open their eyes to new possibilities, and weak signal detection.

How Widely and Well Do We Network?

Vigilant leaders see sooner by acting like explorers. They participate in diverse and unconventional knowledge networks to expand their worldviews. Engaging in diverse external webs, in which weak signals may be serendipitous, is especially challenging for operationally focused managers who have already much on their plates. They tend to be more comfortable within the evident relevance of clearly defined and long-established industry settings. These are too likely to be echo chambers in which like-minded managers seek and get affirmation of their biases and convictions. Connecting with a network of contacts outside familiar domains requires a tolerance for ambiguity and disagreement, plus some courage and a willingness to take some bets that may not pay off.

For example, look into joining small groups of noncompeting companies such as those inspired by former Medtronic CEO Bill George, called True North Groups. By sharing failures and misses, executives turn their blind spots into building blocks for sharper anticipation. Reflect on how well you create and mine your own networks. What purpose does each one serve? Do they corroborate or confront your beliefs? How can you push yourself further out on the periphery? To scan wider, you might want to start tracking cutting-edge blogs, tap into the wisdom of the crowd, and join LinkedIn interest groups outside your fields. Encouraging diverse views and networking outside someone's comfort zone will cultivate curiosity at many levels. Once curiosity is triggered, employees will think more deeply about weak signals, share their concerns, and help their firms adapt to uncertainty. Yet surprisingly, research shows that most leaders largely suppress curiosity, fearing that it will reduce efficiency or perhaps call into question the strategic plans they are pursuing.[9]

What Are the Stories about Vigilance We Tell around Here?

Examining organizational narratives is a proven way to diagnose the perceived realities within a company.[10] Narratives are much more powerful than facts in getting attention and motivating either action or resistance. Comparing stories about successes from seeing sooner or responding to unwelcome surprises will emphasize the values and beliefs that underpin vigilant behavior. A newly appointed Ford CEO did this during his weekly leadership meetings by asking his leadership team members about anomalies they had recently observed. This intervention soon changed the tone of the meetings, shifting the narrative from operational questions to expressing the kind of curiosity that is so essential to vigilance.

Within vulnerable firms, the prevailing narrative is too often self-defeating. We will hear middle-level managers say, "There are no carrots when it comes to sharing bad news, only sticks.... We tend to shoot the messenger.... We just don't have time to listen and reflect." Another pernicious belief is, "I don't think they really want to hear this," especially after bad news or awkward information has been muted, bit by bit, as it moves up the management chain. As one manager ruefully reflected, "In our culture, only good news can go upward, so it doesn't matter if the CEO supports this. Concerns will be filtered out before she hears them."

When the prevailing narrative is an impediment, all is not lost. Leaders can begin to change the collective mindset by envisioning a desired future state in which the company becomes more vigilant to ride waves of change and shape events rather than reacting slowly. Answering the following questions is a good place to start: If our organization was to act out a narrative about winning with vigilance, what behaviors would we see? What would a storyboard mapping out the way we practice vigilance look like? Once enough people embrace an uplifting and promising narrative, leaders can turn then turn their attention to actually realizing it.

Have We Fully Engaged the Board of Directors?

Too many talented boards simply review and approve management's proposed strategy. Their meeting time is mostly spent on progress reports, compliance problems, and operating issues. Bringing them into the challenges of seeing sooner offers many benefits. Board members can tap into a much wider network of information sources about trends, weak signals, and vigilance practices in leading firms. Boards also have a longer time horizon than the leadership team because management is more incented to deliver short-run results. What most directors desire is more breathing space in board meetings to help position the organization for an uncertain future. Once engaged in being more vigilant, these directors can become "scouts" who are actively listening. At Boeing in years past, for example, the board became concerned with the highly scripted agendas developed by the C-suite and staff, leaving little room to address truly important issues. So the board decided to add a two-hour open time slot to each agenda, titled "things we should really talk about."

With shareholder activism rising and corporate scandals a serious risk, many boards in the United States are moving away from reactive oversight and toward becoming boards that lead.[11] For a compelling example of effectively utilizing the board, consider a leading children's hospital facing an increasingly uncertain future. The board members helped develop alternative scenarios for the coming ten years. Together with the hospital's leadership, they foresaw that the most daunting uncertainties would be the availability of funds and whether the pace of scientific advances in genomics and personalized medicine would evolve slowly or rapidly and disruptively. Different assumptions about these two uncertainties had profound implications for the need to acquire new areas of practice and develop new capabilities. Because board members were involved in constructing the scenarios, they also helped the executive team identify early warning signs and probe their meaning. These activities included

- providing top-level guidance about realistic philanthropic expectations;
- helping assess how to practice medicine remotely, especially teleradiology;
- attracting a few highly regarded venture capital leaders to the advisory board; and
- periodically updating the external scenarios and further stress-testing new initiatives.

What Is Our Vigilance Quotient? Are We Vigilant or Vulnerable?

Stepping up to the implications of the previous five questions means the leadership team will have to allocate most of its scarce attention to vigilance. The vigilance survey in appendix A offers a practical guide to the areas most in need of attention. This survey is especially revealing when first completed by each member of a leadership team individually, to identify areas of agreement as well as differences in the diagnosis. In one vulnerable firm, for example, most leaders agreed that failure was typically viewed as an error—not as a learning opportunity. This prompted soul-searching that led to deliberate efforts to relax the approval practices for innovation projects because they had become overly cautious. The company realized this by conducting postmortems of recently failed innovation efforts and discovering that most actually contained key lessons. Armed with a better understanding of the company's zones of vulnerability, leadership changed its approach to innovation—making more room for well-intentioned failures and learning how to see the silver linings in each.

Step 2: Change How Strategy Is Created

The way strategic choices are made in many firms can actually increase their vulnerability if too few people outside the leadership team participate in the process or if cumbersome steps in the process make it a formulaic exercise. The risk is that the firm starts to suffer from an inside-out bias and a near-term orientation. To become more vigilant, such firms need to widen their scope, lengthen the time horizon, and expand the number of people who are involved beyond the leadership team. Some of the features of a vigilant strategy formulation process follow.

Take an outside-in approach. Vigilant firms start their strategy dialogue by discussing what's happening with customers, competitors, and adjacent technologies. This approach was embraced by Mastercard as it focused its strategic thinking on how to compete with cash. It found that many people preferred cash over credit cards because of the personal control. Others were forced by merchants to use cash, and many people in developing economies simply didn't have bank accounts. But Mastercard also found that even the most reluctant adopters of credit cards preferred a cashless solution in such areas as transportation and meals because handling cash often became cumbersome in these settings. For example, when people need to catch a train or grab a meal on the run, they want a quick payment process rather than having to wait in line while they and others find the right cash amount in their wallet or purse. For Mastercard, this deeper insight led to a partnership with Transport for London that resulted in a contactless payment solution, allowing travelers to skip the queues for the tube or bus.

Balance formal and emergent planning. Successful strategies emerge from two interlocking processes: upfront planning, followed by learning by doing. The deliberate or intentional part is more structured and based on rigorous analysis of customer requirements, cost competitiveness, technology road maps, drivers of growth, and so on.[12] This disciplined approach has to be balanced, however, with an emergent and opportunistic process of trial and error learning that allows for exploration and for midstream adjustments. The emergent part entails much experimentation, especially when innovations are involved. Those tests that succeed should be then be acted upon quickly, whereas the disappointments should be mined for their learning opportunities. By embracing this dual approach, outside-in firms learn to excel at both the formal and the emergent aspects of strategy, letting resources flow to the best opportunities while quickly neutralizing threats.

Embrace anomalies. Emergent strategy processes emphasize surfacing anomalies and then addressing them while a strategy is being formed.[13] Whereas vulnerable organizations ignore information that doesn't fit or convince themselves that it isn't important, vigilant firms *seek out* anomalies as early warning signs. Intuit calls this approach *savoring the surprise.*[14] Once its leadership team members saw that some users of the online money management service Mint weren't behaving the way the young-professional target market was "supposed" to behave, they dug deeper and found that these users had adopted Mint to manage their self-employment income and spending. Many, it turned out, were Uber or Lyft drivers, operating in the expanding gig economy. Embracing this market insight, Intuit designed a variation of QuickBooks especially for self-employed workers—which became its fastest-growing product. This novel product extension opportunity would never have been surfaced if Intuit hadn't been studying its customers rigorously with an eye out for anomalies.

Answer guiding questions. We recommend an issues-based strategy process focused on addressing those big questions whose possible answers will shape future results. One of the best sources of these pivotal issues is the set of guiding questions in chapter 3 for systematically learning from the past, interrogating the present, and anticipating the future. These questions serve as an antidote to the precise and targeted focal questions used to evaluate and improve current operations. Vulnerable organizations usually are intent on asking and answering questions about capability utilization, cost variances, and market share changes, but often at the expense of seeing the big picture. The risk is that they see far too many trees and not enough of the overall forest.

Step 3: Invest in Foresight

The strongest signal of leadership commitment is putting serious resources into surveillance, scanning, and other activities aimed at improving vigilance. This shifts the narrative from operational concerns and current performance to looking for what may be coming over the horizon. In most companies, leadership conversations are about how much to budget for R&D or how to balance the project portfolio. But at W. L. Gore and Associates, the $3.1 billion materials technology firm best known for its rain-resistant Gore-Tex fabrics, deeper conversations are encouraged. For example, CEO Terri Kelly wants to know, "How do we create the right conditions where collaboration and sharing occurs naturally … where people want to work together … to be part of something greater than the individual contribution?"

Investments to enhance vigilance through better foresight start by strengthening the ability of the organization to anticipate surprises and take precautions against unexpected events. Spotting weak signals should be everyone's concern but often is no one's responsibility. One way to fix that problem is with a small foresight unit or team that can (1) scour the technical literature, patent filings, and news feeds for early warning signs; (2) send out scouting teams to precursor markets or innovation hubs such as Silicon Valley; (3) engage with venture teams to learn about new relevant investments; (4) mobilize "red teams" to role-play competitors' moves; or (5) conduct low-cost experiments to probe critical uncertainties and capture their lessons.

When Ajay Banga challenged different groups throughout his organization to allocate budget and energy to finding new opportunities, he sent a very strong message throughout Mastercard. He also created Mastercard Labs, with a mandate to generate disruptive services, and then had Mastercard Labs report directly to him as a further sign of strong commitment. The labs operated as a central hub responsible for nurturing a culture of innovation, generating and capturing good ideas from inside and outside the firm. To stimulate broad and novel ideas, the labs organized events such as Take Initiative, a two-day hackathon for developing and testing ideas around a specific challenge. When such commitments are made by top leaders,[15] they bind the organization to a future direction and set the right tone at the top. The commitments of vigilant leaders like Banga aim to introduce a shared narrative about the importance of taking a long view and acting proactively.

Other ways to invest in foresight are to mount a disciplined search for threats and opportunities and engage with an ecosystem of partners. Let's look at each approach.

Mount a disciplined search. Few firms do this better than Intuit. When its leadership in 2010 first saw the possibilities of adapting personal finance solutions to mobile devices, there was resistance from managers who were sure that there was no money to be made with mobile. So Intuit found companies making solid profits from mobile applications and assigned some of its own managers to interview their executives. These contrarian interviews were then shared at an off-site event. As Scott Cook, the founder and chairman of Intuit, recalled, "We didn't have enough time in the meeting.... They wouldn't stop talking. They were convincing each other there was money to be made in mobile."[16] Today, with Intuit's TurboTax mobile app, it's possible to prepare and file a full 1040 form from your phone.

In the same spirit, Walmart, the exemplar of a sclerotic, established retailer, has embraced a disciplined search with an incubator called Store No. 8 (named after an early store where founder Sam Walton conducted business experiments).[17] This incubator is tasked with imagining "new verticals, capabilities, and existential threats" that Walmart might face in the next decade so that it can develop strategies to stay ahead of competitors—especially Amazon. The businesses that are incubated will be run as start-ups, with Walmart investing in them like a venture capitalist and then growing the group as a portfolio. One start-up that Store No. 8 acquired is exploring how virtual reality can augment customers' shopping experiences. They forecast that within a decade, many households will have a super-high-resolution and streamlined VR headset that may reinvent merchandising.

Engage partners in the ecosystem. Mastercard has adopted this approach in recognition that the best opportunities often lie at the intersections where market spaces and industries overlap. Thus, it worked with Maytag to develop Clothespin, a wireless laundry equipment pay-per-use solution that connects washers and dryers to smartphones and laundry equipment service providers. These uncommon partners drew on available technologies—smartphones, online credit card payment systems, and devices connected through wireless cloud communications—to devise a novel solution to make doing laundry away from home easier.

This ecosystem engagement approach was then codified into a platform called Ecosystem Design and Development (EDD),[18] allowing Mastercard to move into many new domains. The platform was managed by Tara Nathan, executive vice president of public-private partnerships, who compares ecosystems to ten-sided networks—in contrast to the classic two-sided network—highlighting the coordination challenges of multiparty innovation webs. The EDD platform enabled Mastercard to collaborate with NGOs in the private sector and with United Nations agencies to help beneficiaries in developing countries receive social services from multiple sources through a single ID. This enabled a secure exchange of data and transparency into the delivery of benefits while reducing leakage and fraud. The process was further digitized by modifying Mastercard's prepaid card infrastructure.

Step 4: Align the Organization

When drafting the Declaration of Independence in 1776, Ben Franklin famously said, "Surely we must all hang together lest we all hang separately" (as traitors to the British Crown). When leaders bring new strategies to life, the biggest obstacle they face is bringing others along the desired path. The gap between strategy and execution remains devilishly hard to bridge.

Getting alignment on the need for heightened vigilance and agreement on what has to be done is an affirmation of the action agenda that puts the organizational pieces in place. This is especially daunting for *vulnerable organizations* in which most managers are comfortable with their current performance but leadership has become anxious about the future. They see digital turbulences looming, doubt the sustainability of the current business model, and are starting to see things happening in the market that can't be readily explained.

Closer alignment of levels and functions means convincing the broader organization of the need for action. There will be pushback and obstacles to overcome, such as, "We are doing this already … we barely have time to do our day jobs and just don't have the extra capacity." The case for vigilance can be made by revisiting past surprises that should have been caught earlier and assessing the cost and waste from being forced to react to competitive moves or the damage sustained by overlooking or suppressing internal threats. Reminders of the long-run damage to Volkswagen from its defeat device debacle or Dansk Bank being oblivious to unusual bank transfers from Russia, are salutary. Leaders also should make clear that vigilance is a team sport and that wearing functional hats requires listening to the concerns of others.

The inability to align stakeholders to strategy implementation takes a heavy toll on leader effectiveness. Efficiency, productivity, and spirit are drained when needed most. The solution requires a human touch that connects with people's interests and their natural desires to work effectively in teams. Here are some proven ways:[19]

1. *Communicate your intent early, often, and simply.* According to Chip and Dan Heath, "What looks like resistance is often a lack of clarity."[20] Leaders can get so invested in their strategic agenda that they fail to realize that what's clear to them is fuzzy to others. Without an abundance of clear communication about where you're going and why, team members may default to behaviors that inadvertently undermine the strategic intent.
2. *Reach out to those who have a stake in your direction.* It was easier for leaders of command and control organizations to line up the troops and send them marching. But organizations have moved from hierarchical structures with clear lines of authority to horizontal networks in which decision-making is more diffuse. These organizations should especially try to include all those with a significant stake in the outcome. Wider communication may slow things down at first, but you will easily make up that time by having better implementation and execution.
3. *Promote open dialogue and true debate.* One of the oldest axioms for change leaders is to *approach* resistance, not ignore it or fight it. Yet too often we fail to even surface, let alone fully understand, why people see things differently. This failure readily occurs when people have their heads down and are focused on immediate demands. And yes, it is messy and often awkward to surface and deal with conflicting views. Yet we pay a heavy price when we don't attend to differences up front. It takes courage to face conflict, tease out diverse expectations, and manage differences.
4. *Reward those who truly take ownership.* Strategic change can only be achieved when people own the solutions in their guts, not just their heads. What we may think is buy-in is often superficial agreement. When a client asked us to help translate their strategy into action, the first thing we did was a "lessons learned" debrief on why they failed to execute successfully before. It became apparent that many

leaders did not share the same urgency or ownership of the prior initiatives, even those assigned to them. This is common when leaders fail to clarify their change agenda, connect key stakeholders, and promote debate of the issues. Rallying people around an execution strategy is hard work. Unless you rely on a proven approach to align stakeholders, you may have no more than an illusion of agreement.

A common barrier to vigilance taking root deep inside the organization is the silent middle that remains unconvinced or confused about what is expected.[21] Here are some of the problems leaders may have to overcome: "This will cause a flood of false signals that will clog everything ... the money would be better spent on getting our new product to market faster ... marketing and R&D will be upset their role is being taken out of their hands." Such objections reflect a mindset that activities and initiatives should have a clear, calculable payoff. It assumes that the ordinary course of business matters more day to day than investing in dynamic capabilities. The counter-response from leadership should be that vigilance is about assuring a healthy future in a changing world. Ideally, tensions between getting bread on the table now versus assuring survival later would have been addressed in the strategic planning phase. Without a clear vision and sustained commitment by leaders to institutionalize organizational vigilance, alignment conflicts usually will be resolved in favor of the status quo.

A recurring theme of successful alignment initiatives is the efficacy of small, empowered teams. As Jeff Bezos once said, "No team should be any bigger than can be fed on two pizzas." At Intuit, new ideas are initially developed by small "discovery teams" that don't report up the chain of command but instead have a direct line to the divisional general manager. Such small teams are especially suited to enhancing vigilance. Recently, seventy Intuit senior managers were grouped into small teams and told to investigate eight major trends that had surfaced either in customer interactions or diverse technology forums (such as how kids under ten years of age use technology, conversational user interfaces, and blockchains). Their reports sparked fifteen further topics, all of which were investigated with hundreds of customer observation studies, interviews with thought leaders ranging from Marc Andreesen to the founders of Airbnb and Uber, and field experiments in diverse geographic markets.

More fundamental organizational transformations to heighten vigilance may have to include structural separation as well. This allows exploration and exploitation activities to occur in newly created units that follow different procedures, reporting rules, and incentives as needed. By putting all sensing and experimentation activities together in a single unit, with a supportive team that relies less on hierarchy and more on behavioral integration, better alignment can be achieved.[22] This approach should be considered whenever the dynamic capabilities needed for greater vigilance conflict with the current business model and operational demands.

8 Forewarned Is Forearmed: Six Lessons about Vigilance

> The greatest danger in times of turbulence is not the turbulence; it is to act with yesterday's logic.
>
> —Peter F. Drucker[1]

When turbulence is pervasive, familiar logics and approaches are obstacles, and improving vigilance becomes a survival mechanism. Several decades after the Internet turned commercial, the pace of digital transformation continues to accelerate. Organizations that thrive when there is greater digital turbulence and upheaval don't wait for the fog of uncertainty about the consequences to lift. Instead, their leadership team has nurtured superior vigilance capabilities for anticipating potential threats, spotting latent opportunities, and acting faster when the time is right.

Vigilance can be systematically developed and strengthened by a motivated leadership team. The underlying capabilities need constant renewing, as vigilant firms can easily lose their edge when short-run operational issues absorb leaders' attention. The reward comes when challenges are thwarted or turned into advantages, and the organization can act confidently without being bewildered by escalating uncertainty. These leadership themes underlie the following six lessons from our book for successfully navigating digital turbulence:

1. Vigilance is achieved through deep curiosity and acute situational awareness guided by insightful questions.

2. Leaders demonstrate their commitment to vigilance by stimulating discovery and debate, encouraging the surfacing of weak signals, and thinking expansively.
3. Vigilance is woven into organizations through investments in foresight capabilities that enable agility
4. Noticing a signal is not the same as understanding what it means or knowing how to act wisely in view of the broader context and after connecting the dots.
5. The collective attention of a leadership team is a scarce and valuable resource that can be squandered easily.
6. Incumbents can fight back! Digital turbulence is not necessarily disruptive.

1: Vigilance Is Achieved through Deep Curiosity and Acute Situational Awareness Guided by Insightful Questions.

Weak signals from the fuzzy zone on the external periphery or deep within the organization are faint and blurry when first spotted. By the time a clear prediction or forecast can be made, it is usually too late; others will have already acted, leaving the vulnerable firm no choice but to react and adopt a defensive crouch. The challenge is that weak signals are surrounded and distorted by distracting noise.

The issue is not a lack of data, but a lack of good questions to guide where to look while filtering out the distractions. Managers may console themselves by collecting more data, but unless this effort reaches outside familiar boundaries, they may miss looming threats or latent opportunities. A balance must be found between undirected search, like that of a lighthouse that scans the horizon 360 degrees every ten minutes, and a myopically focused search that monitors a tiny area within a vast domain. The aim is to create situational awareness within the executive team about how the world is changing and why, and then to explore these changes.

The best guiding questions are thought starters for probing places that might escape the attention of operationally inclined managers.

We posed three types of guiding questions in chapter 4. The first type of question aims to *learn from the past* by more deeply understanding and reducing past systemic vulnerabilities. The second type of question *interrogates the present.* Most surprises have antecedents that first appear as anomalies that are outside what is normal or expected. The third type of question *anticipates the future,* with mind-expanding scenarios, red team role-playing, or imagining a future in which a weak signal has been pushed to an extreme. For example, "What will happen if non-bank digital technology firms such as Google or Alibaba seriously enter payment services or grab a major market share in lending with a new suite of digital services?" These guiding questions sensitize leaders to changes that are just faintly visible around the corner.

Some of the turbulence that shook Facebook in 2018 came from collectively ignoring this lesson, by not paying attention to anomalies while dismissing red flags. They were slow to question the spread of disinformation or manipulative campaign messages and did not seem to see a downside to giving third parties access to user information.

2: Leaders Demonstrate Their Commitment to Vigilance by Stimulating Discovery and Debate, Encouraging the Surfacing of Weak Signals, and Thinking Expansively.

When firms are surprised, someone within the organization or the network of partners probably knew about it earlier. But the leadership team didn't know they knew—and the frontline people didn't know the leaders needed to know. This failure is a sign that leaders have not made vigilance a sufficient priority. At the best of times, leaders receive only a fraction of the information they need, and what they receive may be slanted to please them or serve hidden agendas.[2] In turbulent times, upward communication channels need to become wider and faster. Advances in digital technologies should be leveraged to guard against information overload and help leaders with signal detection. Well-known examples where leaders failed in this regard include Volkswagen's use of defeat devices to skirt emissions regulations, Wells Fargo's aggressive or illegal sales tactics, and Dansk Bank's involvement in a massive money laundering scandal. Other striking cases include Theranos, founded in 2003 to simplify blood tests, which closed its doors in 2018 with a record loss to investors of $1 billion, and the still-unfolding Purdue Pharma scandal about pushing unconscionably high levels of opioids.

Vigilant leadership is about taking the long view, anticipating new patterns forming, and taking sensible risks when acting on incomplete information. Employees will actively scout the periphery or share concerns about festering internal threats when the senior command endorses and rewards such excursions. Leaders set the tone at the top and further enforce their commitment to vigilance with visible investments in foresight activities. The outsize influence of leadership was underscored by a very successful private equity investor, who wanted to invest in firms well attuned to their environment. As a result, the companies in his portfolio could see possibilities and threats sooner than their rivals. He noted that "those leaders that are tightly focused on the existing operations are good COOs but they are not leaders for the long haul." As he reflected further, he realized that successful CEOs were secure and had no hesitation in surrounding themselves with strong people. They encouraged vigorous debate and dialogue because they didn't need to pretend to have all the answers. He summed up by observing: "If you only follow the conventional and obvious implications, you'll just be part of the crowd. Success comes from thinking more imaginatively."

These attributes of vigilant leaders are rewarded in all organizations. Consider the words of Dr. Stephen Klasko, CEO of the Jefferson Health System, who has led the reimagining of health care in a 195-year-old academic medical center:[3] "We decided to go all in on telehealth in 2013, and people thought I was on drugs. I wanted to be the only telehealth program in the country fully staffed by our ER clinicians, so every doctor on our telehealth platform would be a Jefferson doctor, which we did by aligning incentives and goals." Thanks to many such initiatives, Jefferson has become a leader in digital innovations of the patient experience.

Lesson 2 was seemingly not absorbed by Facebook leadership when Alex Stamos (head of cybersecurity) warned about Russia's weaponization of Facebook's platform. Frustrated with CEO Mark Zuckerberg's and COO Sheryl Sandberg's low levels of concern, Stamos decided to bypass them and warn board members as a last resort. His alert prompted a board grilling of leadership and caused Sandberg afterward to yell at Stamos, "You threw us under the bus." Sandberg had badly misread the growing problems festering on her watch. [4] Her desire to protect Facebook caused her to back away from the gathering storms about hate speech, bullying, and other toxic content on their platform. And Zuckerberg often challenged internal dissent and seemed to favor loyalty over debate. As Elliot Schrage, VP of global communications, marketing, and public policy, said, "We failed to look and try to imagine what was hiding behind the corners."

3: Vigilance Is Woven into Organizations through Investments in Foresight Capabilities That Enable Agility.

Superior vigilance is the product of a web of dynamic capabilities enabling firms to thrive during turbulence by *sensing* opportunities and threats sooner than their rivals do and then *seizing* the best opportunities while confronting the threats with prudent preparations and continuous learning. This gives the leadership team confidence to act when uncertainty abates. These two sets of dynamic capabilities reinforce each other. Seeing sooner alerts the organization to probe further and be prepared to act faster when the time is right.[5] This is the essence of agility.

A third capability is about continuous organizational transformation that sustains an agile way of operating. To reap the benefits of rapid, iterative teamwork, the organization needs to layer this *transforming* capability on top of the sensing and seizing ones. Taken together, these three types of capabilities create an organization with foresight that is loosely coupled and open rather than rigid, controlling, and hierarchical. These dynamic capabilities need to be embedded within a culture that reinforces an outside-in entrepreneurial mindset. The motivating force that shapes this culture and invests in these capabilities is a leadership team committed to becoming and staying vigilant.

Each sensing, seizing, and transforming dynamic capability has numerous supporting capabilities that are orchestrated by the leadership team. Superior sensing delivers an insight advantage through a combination of probe-and-learn inquiry systems, competitive intelligence gathering, deep data analytics, and rapid experimentation that tests hypotheses about the causes and consequences of weak signals. What a financial firm needs to master in its insights toolbox, however, will be different from a pharmaceutical firm navigating a complex and opaque health system. The strategy and capabilities that underpin organizational vigilance need to be customized for each business unit and adjusted over time.

Facebook also failed to absorb the implications of lesson 3. Even though Facebook's data science experts excelled at using deep learning and predictive analytics to glean as much as possible about their users, the company failed to turn its foresight acumen on itself and remained a prisoner of its blind spots. When criticized by outsiders, Facebook was notably slow in grasping the deeper meanings of these criticisms and failed to correct its own organizational vulnerabilities in time. The company did not draw sufficiently on the collective wisdom of the entire enterprise and beyond.

4: Noticing a Signal Is Not the Same as Comprehending What It Means or Knowing How to Act Wisely.

Leaders must first confirm an apparently ambiguous signal by looking at it from different angles and connecting it with other signals. This triangulation is best done by bringing different people with diverse views and experiences into the process. The inevitable and healthy differences in viewpoints will surface competing hypotheses and illuminate diverse parts of an incomplete picture. This will help the team think creatively and "connect the dots" better. Being objective and dispassionate about changing external realities is hard when it calls into question the current strategy or deeply held beliefs and assumptions. Managers and leaders are seldom dispassionate bystanders, trying to understand an unfolding situation. They tend to be deeply invested participants whose hopes and fears, as well as their reputations, compensation, and careers, shape their sense making and preferences about competing courses of action.

Executives are often torn between keeping morale high and fully acknowledging bad news from the field. When managers in GE's power business became justly worried in late 2016 about very rosy revenue projections for gas turbines (the titanic machines for power plants), they told Paul McElhinney—the head of the unit—of their skepticism. He reported directly to Steve Bolze, who was the head of GE Power and widely considered to be next in line to become CEO of GE. Instead of treating these concerns seriously as worrisome signals about problems lurking around the corner, McElhinney stopped them cold by saying, "Steve's our guy. Get on board; we have to make the numbers."[6] A year later, profits cratered and the reality of the dire situation reached the board, which then fired most of the leadership team. Clearly, leaders need to pay attention to where and why their on-the-ground employees are pushing back because such pushback may be a warning sign that the higher-up leaders are out of touch.

Effective leaders escape these traps not just by relying on their direct reports but by reaching deep into the organization while also leveraging their external network as a personal radar. In fast-moving situations, it is imperative that highly confident leaders first scan widely and learn before taking a strong position. The essence of seeing sooner is to act wisely, informed by probing inquiries followed by quick reality checks to assess competing hypotheses. Such interactive learning loops allow them to invest in a portfolio of real options that maintain strategic flexibility where needed. The aim is to engage in calculated risk-taking guided by the ethical norms of stakeholders and society at large. Through faster exploratory learning, a firm can avoid having to react in haste once the fog has lifted and after its degrees of freedom have been narrowed or fully preempted by more agile rivals.

Ignoring the message of lesson 4 further hampered Facebook because its options became limited after a slow response to mounting criticisms about its business model and inadequate protection of private user information. In this digital age, however, the strategic issues go far beyond privacy and touch the legitimacy of the business models used by social media companies. They concern issues of corporate governance, market power, government control, the monetization of individual data, and honoring implicit social contracts about fairness and good citizenship. These deeper concerns apparently did not get much play at Facebook. Roger McNamee, who advised Facebook early on, said the company deployed "the most centralized decision-making structure I have ever encountered in a large company."[7]

5: The Collective Attention of a Leadership Team Is a Scarce and Valuable Resource and Can Be Squandered Easily.

"Pay attention" is a helpful dictum for parents with distracted children or leaders of organizations overloaded with weak signals of possible threats and opportunities. Harried leaders must be strategic about how their collective attention resource is managed. It can be *expanded* by reducing the need to process or by increasing the capacity for processing information. A senior leader's full plate can often be freed up through better prioritization and delegation of activities. Leaders can also increase their capacity to process information through the use of digital search and interpretation technologies and by assigning greater accountability for "collecting the paranoia." Also, their attention can be *more wisely allocated* by learning from past surprises, using digital technologies adroitly, and being sensitized to deep-seated cognitive biases.

Individual attention is a fixed resource at any given point in time. But collective attention can be enhanced or diminished through setting priorities, changing incentives, and deploying time-saving technologies for routine tasks. With continued advances in AI, overwhelmed leaders can reduce their administrative load and create slack for strategic thinking and peripheral scanning.

Collective attention can easily be subverted, however, by such organizational maladies as tunnel vision, short-termism, strategic gaming, and wishful thinking. The intent in these cases need not be nefarious since subordinates may fairly ask for a share of their leaders' limited reservoir of time. Most employees cannot reasonably know a leader's capacity to handle more issues or tasks, which makes it essential for leaders to be strategic in their allocation of attention. In hindsight, Facebook leadership's limited attention was focused on the wrong issues, such as technological fixes for social issues complex in scope and nature. Zuckerberg realized belatedly that Facebook's deeper challenges are about *individual privacy and security of personhood.*

An especially pernicious malady in organizations is willful blindness,[8] which may seem oxymoronic at first. How can one be aware and then pretend not to be aware? Organizations may turn a blind eye to flashing warning signs of lurking problems, to stave off unwanted conflicts, protect economic interests or power, or just reduce anxiety. But willful blindness can easily turn issues into ticking time bombs, as Volkswagen, Wells Fargo, Dansk Bank, or Facebook have learned to their detriment. To avoid such toxic consequences, leaders must encourage diversity and debate and not let deep-seated challenges continue to fester. Effective leaders know when the brain of the firm is registering problems in one part of the organization while also willfully ignoring the same issues elsewhere. Vigilant organizations outperform their vulnerable rivals by taking decisive remedial actions before critical issues spiral out of control. Such timely actions also make it possible for an organization to avoid having to respond defensively with only limited options and paying a steep price for reacting late.

6: Incumbents Can Fight Back! Digital Turbulence Is Not Necessarily Disruptive.

Few management concepts in the digital age have had as much resonance and influence as the theory of disruptive innovation.[9] The basic message is a stark warning for incumbents who are vulnerable to attacks by disruptive entrants with robust technology or business model advantages that can quickly scale up in search of new customers. The theory also assumes that the hapless incumbents would keep improving their performance at a pace that overshot the needs of most customers. For example, the personal computer disrupted low-end mini-computers relying on costly custom systems because they could sustain the radical cost advantages gained by using standard components as they grew. Another assumption of the theory was that incumbents would be reluctant to respond to disrupters due to their fear of cannibalizing their cash-generating core business.

Ironically, the heightened awareness of the threat of disruption has considerably reduced the likelihood of incumbents being blindsided this way. The theory still provides a warning of what might happen if the forces behind digital turbulence are ignored. But this is not a substitute for thoughtful foresight that converts early warning signs of disruptions into initiatives that can be creatively tested. These learning probes are usually followed by "small bets" on start-ups, anticipatory development projects, and joint ventures that can be unwound if things don't work out as expected. Many big firms are buying start-up challengers, just as Walmart acquired Jet.com and UPS bought Coyote Logistics, which aspired to be the Uber of trucking. The aim of these moves is to prepare the organization to act faster.

Vigilant incumbents are increasingly heeding warnings about the consequences of inaction or a cautious wait-and-see posture. They have honed their skills to acquire embryonic disruptors and thus absorb their digital capabilities and scarce talent. By seeing the threat of disruption sooner, they are quicker to appreciate which digital skills are going to be in short supply and can move earlier than their rivals to invest in digital talent recruitment, development, and retention. The success of Adobe Inc. as featured in chapter 2 depended on acting preemptively to control its own destiny. The message for vigilant firms is to be alert to technology advances that potentially offer greater customer value and not wait for an outsider to exploit it first. Watch novel upstarts carefully from the outset, learn from them, and treat the potential disruption as opportunity. Forewarned is indeed forearmed.

Closing Message

Just as awareness that a sudden outflow of the tide might signal an incoming tsunami, recognizing the early signals of threats or opportunities has become a priority for all firms. The leadership challenge is increasingly about creating new hands of cards—not about refining how to play an existing hand. The future game in business will be one of agility and adaptation—not as a lone ranger, but as a valued member of external networks and ecosystems that leverage a highly interconnected world for joint gains. Organizations doing this well have a great advantage. As Charles Darwin observed, "It's not the strongest of the species who survive, nor the most intelligent, but the ones most responsive to change."[10] We equate "responsiveness to change" with being a vigilant organization. It starts with a diverse, concerned, committed, open-minded, and curious leadership team sharing a forward-looking vision.

The path to vigilance provided in this book has no ending. Leaders must work tirelessly to keep their organizations at a heightened state of alertness and poised to act on early signals from their inside and outside environments. There is no room for complacency if they are to stay ahead of digital turbulence and as yet barely imagined future scenarios. The rewards for vigilant organizations include stronger market positions, higher profits and growth, more motivated employees, and longer life expectancies.

Appendix A A Diagnostic Tool to Assess Vigilance

How Well Does Your Organization See Sooner and Act Faster?

The survey presented here is designed to assess *your organization's* capacity to *sense and act on weak signals of threats and opportunities*, both inside and outside the enterprise. By *your organization* we mean a business unit, division, or the entire enterprise. Please clarify which one you want to assess.

When used in a team setting, start with each person completing this assessment tool individually and then submitting the results, either anonymously or with their name included. Once the combined results have been summarized, focus on those items for which either the mean scores are low or there is considerable variance in the assessment scores. Both are good areas to discuss with the whole group in term of perspectives, examples, causes and remedies.

A. How Well Does Your Organization Sense and Respond to Weak Signals?

1. In the past three years, how often have you been surprised by a **threat from outside** your company (e.g., new entrants, technologies, regulations, business models, etc.)?

Frequently (compared to our peers)	☐ 1	☐ 2	☐ 3	☐ 4	☐ 5	☐ 6	☐ 7	Seldom (compared to our peers and rivals)

2. In the past three years, how often have you been surprised by a **threat from inside** your company (e.g., fraud, harassment, malfeasance, incompetence, or other bad behavior)?

Frequently (compared to peer/ rivals)	☐ 1	☐ 2	☐ 3	☐ 4	☐ 5	☐ 6	☐ 7	Seldom (compared to our peers and rivals)

3. How good are you at spotting **opportunities originating outside** your company (e.g., new customers, suppliers, partners, business models, channels, technologies, social trends)?

We are usually late and often have to catch up	☐ 1	☐ 2	☐ 3	☐ 4	☐ 5	☐ 6	☐ 7	We often see opportunities well ahead of our key rivals

4. How good are you at spotting **opportunities inside your company** (e.g., emerging talent, innovative ideas, efficiency improvements, better incentive systems, IT applications)?

We often overlook good ideas right under our noses	☐ 1	☐ 2	☐ 3	☐ 4	☐ 5	☐ 6	☐ 7	We scan for and develop new ideas from inside vigorously

5. How quickly have you detected **fundamental shifts** in your industry (e.g., changes in business models, competitors, disruptive technology, new regulations, or recessions)?

Laggards: We are always slow or late in seeing shifts	☐ 1	☐ 2	☐ 3	☐ 4	☐ 5	☐ 6	☐ 7	Leaders: We effectively anticipated major shifts

6. How accurately have you forecasted relevant developments over the past five years?

Very poorly: We did not try or actual results differed greatly from forecasts	☐ 1	☐ 2	☐ 3	☐ 4	☐ 5	☐ 6	☐ 7	Very well: Actuals have deviated little from our forecasts

7. How extensively does your organization leverage external networks and partnerships in order to see external changes sooner than rivals do?

Seldom used	☐ 1	☐ 2	☐ 3	☐ 4	☐ 5	☐ 6	☐ 7	Widely used

8. How open is your organization to listening to reports from scouts or others on the periphery of your market (i.e., beyond your core business)?

Closed: Our culture discourages listening	☐ 1	☐ 2	☐ 3	☐ 4	☐ 5	☐ 6	☐ 7	Open: Our culture greatly encourages such listening

9. How willing are your frontline employees to forward signals of opportunity or concern **upward** to senior management?

Reluctantly: We lack channels, incentives, or sufficient trust to do this	☐ ☐ ☐ ☐ ☐ ☐ ☐ 1 2 3 4 5 6 7	Eagerly: We have enough trust, recognition, and incentives for doing so

10. How readily is relevant information about the periphery shared **sideways**— across either functions or business units laterally?

Poorly: People hoard or ignore relevant information	☐ ☐ ☐ ☐ ☐ ☐ ☐ 1 2 3 4 5 6 7	Excellently: People share information regularly across silos or boundaries

11. How well does your organization anticipate external changes in the market place compared to your rivals?

Very poorly: We seldom see significant market changes before they do	☐ ☐ ☐ ☐ ☐ ☐ ☐ 1 2 3 4 5 6 7	Excellently: We often see important changes well before our rivals do

12. How involved is your board of directors in spotting issues of concern early and clearly?

Very little: They seldom see significant market changes before we do	☐ ☐ ☐ ☐ ☐ ☐ ☐ 1 2 3 4 5 6 7	Highly: They often raise important issues and help us understand them better sooner

B. What Is the Ability of Your Organization to Respond to Weak Signals?

1. How much time and resources does the leadership team devote to scanning for weak signals of threats and opportunities?

Low priority: Few people actively watch for or examine weak signals	☐ ☐ ☐ ☐ ☐ ☐ ☐ 1 2 3 4 5 6 7	High priority: Many managers actively scan the periphery and explore

2. How willing are senior managers to challenge basic assumptions about your current business model or views about the future?

Closed: Mostly defensive of assumptions or even hostile toward change	☐ ☐ ☐ ☐ ☐ ☐ ☐ 1 2 3 4 5 6 7	Open: Leaders openly encourage challenging assumptions and traditions

3. How far does your organization's strategic planning cycle look into the future?

	1	2	3	4	5	6	7	
Emphasis on short term (two years or fewer)	☐	☐	☐	☐	☐	☐	☐	Emphasis on long term (five years or more)

4. To what extent does your organization use such tools as scenario planning, real options, and predictive analytics when developing strategies?

	1	2	3	4	5	6	7	
Limited use: We use mostly point forecasts and NPV or ROI analyses	☐	☐	☐	☐	☐	☐	☐	Extensive use: We stress-test our strategies and often stage our investments

5. How forward-looking is your strategic planning process?

	1	2	3	4	5	6	7	
Very little: It is rigid, calendar-driven, and political to get access to resources	☐	☐	☐	☐	☐	☐	☐	Very much: It is flexible, collaborative, issues-driven, and managed in real time

6. How agile is the organization when you need to adjust plans midway?

	1	2	3	4	5	6	7	
Very little: Once plans are set, it is difficult to amend	☐	☐	☐	☐	☐	☐	☐	Very much: It is fluid, responsive to threats and opportunities, and agile

7. Does your organization utilize new technologies for posing and analyzing queries to large databases, from data mining to artificial intelligence (AI), such as IBM's Watson system?

	1	2	3	4	5	6	7	
Archaic: Systems are old and difficult to use; no AI	☐	☐	☐	☐	☐	☐	☐	State of the art: Predictive analytic models and machine learning incorporating AI

8. Is there much accountability in your organization for taking timely action when confronted with ambiguous signals of threats and opportunities?

	1	2	3	4	5	6	7	
No one is clearly responsible	☐	☐	☐	☐	☐	☐	☐	Responsibility is clearly assigned to a project team or leadership team

9. Are there tailored incentives in your organization to reward managers who adopt a wider vision and/or who actively explore, share, and interpret weak signals?

No: There are no special incentives in place for acting on weak signals	☐ ☐ ☐ ☐ ☐ ☐ ☐ 1 2 3 4 5 6 7	Yes: Top management provides direct rewards for scanning and fast follow-up

10. To what extent does your leadership team actively build and participate in external networks?

Limited to narrow and routine industry settings (like conferences or events)	☐ ☐ ☐ ☐ ☐ ☐ ☐ 1 2 3 4 5 6 7	Wide engagement in diverse outside social, civic, and professional networks

11. What is your organization's prevailing attitude toward mistakes or innovation setbacks?

Failure is seen as an error	☐ ☐ ☐ ☐ ☐ ☐ ☐ 1 2 3 4 5 6 7	Failure is seen as a learning opportunity

12. How supportive is your board of directors about taking timely action to position your organization well for future changes or about internal issues that need to be resolved in time?

Very little: Our board seldom drives strategic decisions with urgency	☐ ☐ ☐ ☐ ☐ ☐ ☐ 1 2 3 4 5 6 7	Excellent: Our board is truly proactive and supports bold action when needed

C. How Do You Assess Your Organization's External Business Environment?

1. What has the market growth pattern been in your business over the past five years?

Slow and stable	☐ ☐ ☐ ☐ ☐ ☐ ☐ 1 2 3 4 5 6 7	Rapid and unstable

2. What is the speed and direction of technological change in your business?

Slow and predictable	☐ ☐ ☐ ☐ ☐ ☐ ☐ 1 2 3 4 5 6 7	Fast and unpredictable

3. How predictable are the strategies, actions, and intentions of key competitors, suppliers, partners, and other key stakeholders in your business?

Very predictable	☐ ☐ ☐ ☐ ☐ ☐ ☐ 1 2 3 4 5 6 7	Highly unpredictable

4. How susceptible is your organization and industry to macroeconomic forces?

Low sensitivity to price changes, currencies, business cycles, tariffs, and so on	☐ ☐ ☐ ☐ ☐ ☐ ☐ 1 2 3 4 5 6 7	High sensitivity to prices, currencies, business cycles, tariffs, or political changes

5. How sensitive is your organization and business to social and technological changes?

Low: Mostly gradual change from the past	☐ ☐ ☐ ☐ ☐ ☐ ☐ 1 2 3 4 5 6 7	High: Strong chance of major disruptions and changes in business models

6. How much has your industry been transformed or reshaped during the past five years?

Very little: It looks much as it did five years ago	☐ ☐ ☐ ☐ ☐ ☐ ☐ 1 2 3 4 5 6 7	Very different: There are many new players and business models

7. What is the potential for major disruptions in your business over next five years?

Low: Few surprises expected, and mostly things we can handle	☐ ☐ ☐ ☐ ☐ ☐ ☐ 1 2 3 4 5 6 7	High: Several significant shocks are expected

Appendix B About Our Research

Survey Design

We build on some of the questions from our 2006 survey (which can be found in our *Peripheral Vision* book)[1] and added further questions based on specific hypotheses about variables likely to discriminate between vigilant and vulnerable organizations. These hypotheses were based on additional articles we published since 2006, our consulting experiences, and the relevant management and strategy literatures in general. The survey shown in appendix A is the one we used to collect data with some minor modifications[2] as well as an additional section D to capture demographic and other classification information about the respondents and their companies, as explained further in this appendix.

Data Collection

We used several methods to identify senior executive respondents who would be willing to fill out the survey themselves (rather than delegating the task to their staffs). We desired to get just one completed survey per company from a senior leader able to address the questions. Of the 118 respondents we sampled, 30 percent were recruited directly via organizations and consulting firms with whom we had close prior relationships. The remainder were participants in senior executive programs or by-invitation-only conferences at Wharton and elsewhere. In brief, the profile of these 118 respondents is as follows:

- CEOs or members of C-suites made up 45 percent; 55 percent reported to the C-suite.
- The average size of the companies represented was 22,400 employees, with a median size of 1,700 employees.
- On average, 26 percent of sales came from outside the country where a company's main headquarter was located.
- The survey was skewed toward high performance companies: 43 percent described their business performance relative to their industry in the past five years as being "leaders" and 53 percent scored their company as "average performers." Few were below average.

Estimating the Explanatory Model

Because the survey used seven-point, bipolar scales, we created summed score indices to define our dependent plus four independent variables. The latter corresponded closely to the four vigilance elements described in chapter 2—as follows:

X_1 Leadership posture = Questions B2 + A9 + B6 + B10

X_2 Strategy making = Questions A8 + B5 + B11

X_3 Investments in foresight = Questions A7 + B1 + B4

X_4 Accountability = Questions A10 + B8 + B9

The dependent variable was meant to capture each company's vigilance quotient, which we measured as the summed scores of three variables. Each of these entails a component of organizational vigilance performance compared to rivals, as follows:

Y (Vigilance quotient) = A3 (Seeing outside opportunities)
+ A5 (Detecting fundamental shifts in the industry)
+ A11 (Ability to anticipate external changes in the market)

Exploratory regressions with various control variables found that only company size (measured as the number of employees) had a significant effect on explaining variance in the vigilance quotient (Y). The role of leadership (X1) and investments in foresight (X2) is more significant in larger than smaller companies, but directionally the same. A higher score on these dimensions improves vigilance as explained next.

Explaining the Vigilance Quotient

The standardized coefficients in the following table show by how many standard deviations the vigilance quotient variable will change for each standard deviation change in an independent variable.

	Coefficients	VIF
X_1 Leadership posture	.334*	3.8
X_2 Strategy making	.080	3.9
X_3 Investments in foresight	.241*	2.2
X_4 Accountability and coordination	.060	2.7

R^2 (adjusted) = 0.387
Sample size (N) = 118
*Significance level < 0.02

Colinearity is a problem inherent in most surveys, especially when the questions use similar interval scales. If in addition, all positive anchors lie on the same side of the scale (as in our case), the "halo effect" known as *ecological correlation* may set in. Fortunately, the standard test of multicollinearity called the *variance inflation factor* (VIF)[3] showed tolerable levels for each coefficient in our corporate data set.

We also sampled two other populations to test our model more broadly, including ninety-three American foundations procured via the Council on Foundations in Washington, DC, and 134 credit unions obtained with the help of the Filene Institute in Wisconsin. These cross-validation tests confirmed that our model applies to nonprofit entities as well, attesting to its conceptual robustness across organizational types.[4]

Notes

Introduction

1. Peter F. Drucker, *Managing in Turbulent Times* (New York: Harper and Row, 1980), 2.

2. George S. Day and Paul J. H. Schoemaker, "Adapting to Fast-Changing Markets and Technologies," *California Management Review* 58, no. 3 (Spring 2017): 28–37.

3. Valuable introductions to the role of dynamic capabilities can be found in David J. Teece, "Explicating Dynamic Capabilities: The Nature and Microfoundations of (Sustainable) Enterprise Performance," *Strategic Management Journal* 28 (August 2007): 1319–1350; and Paul J. H. Schoemaker, Sohvi Heaton, and David Teece, "Innovation, Dynamic Capabilities and Leadership," *California Management Review* 61, no. 1 (2018): 15–42.

4. Alastair Gale and Sean McLain, "Companies Everywhere Copied Japanese Manufacturing. Now the Model Is Cracking," *Wall Street Journal*, February 4, 2018, https://www.wsj.com/articles/companies-everywhere-copied-japanese-manufacturing-now-the-model-is-cracking-1517771142.

5. Paul J. H. Schoemaker and Philip E. Tetlock, "Building a More Intelligent Enterprise," *MIT Sloan Management Review* 58, no. 3 (Spring 2017): 28–37.

1 Facing Reality in Real Time

1. This quote is engraved on a statute in Helsinki honoring this former president of Finland, who served from 1946 to 1956.

2. We first presented this concept of vigilance in George S. Day and Paul J. H. Schoemaker, "Are You a Vigilant Leader?," *MIT Sloan Management Review* 40 (Spring 2008): 43–51.

3. Joshua Brusten, "Inside Radio Shack's Collapse," *Business Week*; and "Lessons from RadioShack: To Stay on Top, Figure Out What Got You There," *Knowledge@Wharton*, February 15, 2015, https://knowledge.wharton.upenn.edu/article/where-radioshack-went-wrong/.

4. George S. Day and Paul J. H. Schoemaker, *Peripheral Vision: Detecting the Weak Signals that Can Make or Break Your Company* (Cambridge: Harvard Business School Press, 2006).

5. Mattel has since become aware of AR. See Alex Clark, "Mattel Unveils Plan to Reinvent Company and Deliver Enhanced and Sustainable Growth," Mattel, Inc., June 14, 2017; Michal Lev-Ram, "Can a Tech Makeover Save the Toy Industry?," *Fortune*, September 22, 2017; and Devindra Hardawar, "How Mattel Is Using AR to Let You Preview Hot Wheels Playsets," *Engadget*, February 16, 2018.

6. Juli Clover, "Augmented Reality Teddy Bear from Seedling Now Available Exclusively at Apple Retail Stores," *Mac Rumors*, October 23, 2017.

7. Olivia Solon and Julia Carrie Wong, "Jeff Bezos v the World: Why All Companies Fear Death by Amazon," *Guardian*, April 24, 2018.

8. Nate Silver, *The Signal and the Noise: Why So Many Predictions Fail—But Some Don't* (New York: Penguin Press, 2012), was the first we know to report this startling number. But five years later, the same two-year data explosion statistic still applied according to Bernard Marr, *Data Strategy: How to Profit from a World of Big Data, Analytics and the Internet of Things* (London: Kogan Page, 2017).

9. Gordon E Moore, "Cramming More Components onto Integrated Circuits," *Electronics* 38, no. 8 (April 19, 1965): 114.

10. Neil Gershenfeld and Alan Gershenfeld, *Designing Reality: How to Survive and Thrive in the 3rd Digital Revolution* (New York: Basic Books, 2017).

11. Ajay Agrawal, Joshua Gans, and Avi Goldfarb, *Prediction Machines: The Simple Economics of Artificial Intelligence* (Boston: Harvard Business Review Press, 2018).

12. Jay Barney, "Firm Resources and Sustained Competitive Advantage," *Journal of Management* 17 (1991): 117.

13. Rita Gunther McGrath, *The End of Competitive Advantage: How to Keep Your Strategy Moving as Fast as Your Business* (Cambridge, MA: Harvard Business School Press, 2013).

14. IBM's computer program Big Blue beat Garry Kasporov, then the reigning world champion, in New York City in 1997 under official tournament conditions. DeepMind's neural net program, AlphaGo, beat Lee Sedol, then the top-ranked Go champion, four out of five times during 2016 under strict tournament conditions in Seoul, South Korea.

15. *Wall Street Journal*, September 6, 2018, front-page photo.

16. Eric Almquist John Senior and Nicolas Bloch, "The Elements of Value," *Harvard Business Review*, September 1, 2016.

17. Nate Silver, *The Signal and the Noise*.

18. See Matt Egan, "Wells Fargo Strips EO and 7 Top Execs of 2016 Bonuses," *CNN Business*, March 1, 2017, http://money.cnn.com/2017/03/01/investing/wells-fargo-strips-ceo-bonus-fake-account-scandal/.

19. Chris Arnold, "Who Snatched My Car? Wells Fargo Did," NPR, August 2, 2017, http://www.npr.org/2017/08/02/541182948/who-snatched-my-car-wells-fargo-did.

20. Jennifer A. Kingson and Stacy Cowley, "Wells Fargo Employee Started Blowing the Whistle in 2005," *New York Times*, October 12, 2016.

21. Nick Summers, "Inside Chronicle, Alphabet's Cybersecurity Moonshot," *Engadget*, November 30, 2018.

22. Henrik Legind Larsen, José María Blanco, Raquel Pastor Pastor, and Ronald R. Yager, eds., *Using Open Data to Detect Organized Crime Threats: Factors Driving Future Crime* (Cham, Switzerland: Springer, 2017).

23. Betsy Morris and Deepa Seetharaman, "The New Copycats: How Facebook Squashes Competition from Startups," *Wall Street Journal*, August 9, 2017.

24. Herbert A. Simon, "Designing Organizations for an Information Rich World," in *Computers, Communication and the Public Interest*, ed. Martin Greenberger (Baltimore, MD: The Johns Hopkins Press, 1971), 40–41. For a review of the latest thinking on organizational attention, see William Ocasio, "Attention to Attention," *Organization Science* 22, no. 5 (September/October 2011): 1286–1296.

25. Sarah M. Caldicott, "Why Ford's Alan Mullaly Is an Innovation CEO for the Record Books," *Forbes*, June 25, 2014. For a deeper discussion, also see Harbir Singh and Mike Useem, *The Leader's Checklist* (Philadelphia: Wharton Digital Press, 2017).

26. See "24/7 Wall Street: Interview with Ford CEO Alan Mulally, September 16, 2009, http://www.newsweek.com/247-wall-street-interview-ford-ceo-alan-mulally-79611.

27. George Stalk, *Competing against Time: How Time-Based Competition Is Reshaping Global Markets* (New York: Simon and Schuster, 1990); William R. Tobert, *Action Inquiry: The Secret of Timely and Transforming Leadership* (San Francisco: Berrett-Koehler Publishers, 2004); and Colin Price, *Accelerating Performance* (Hoboken, NJ: Wiley, 2017).

28. David J. Teece, Margaret Peteraf, and Solivi Leiih, "Dynamic Capabilities and Organizational Agility: Risk, Uncertainty and Entrepreneurial Management in the Innovation Economy," *California Management Review* 58, no. 4 (Summer 2016): 13–35.

29. D. J. Teece, "The Foundations of Enterprise Performance: Dynamic and Ordinary Capabilities in an (Economic) Theory of Firms," *Academy of Management Perspectives* 8, no. 4 (2014): 328–352.

30. George S. Day and Paul J. H. Schoemaker, "Adapting to Fast-Changing Markets and Technologies," *California Management Review* 58 (Summer 2016): 59–77.

31. Darrell K. Rigby, Jeff Sutherland, and Hirotako Takeuchi, "Embracing Agile: How to Master the Process That's Transforming Management," *Harvard Business Review* 94, no. 5 (May 2016): 41–50.

32. René Rohrbeck and Menes Etingue Kum, "Corporate Foresight and Its Impact on Firm Performance: A Longitudinal Analysis," *Technological Forecasting and Social Change* 129 (2018): 105–116.

33. Rohrbeck and Kum, "Corporate Foresight and Its Impact on Firm Performance."

34. G. S. Day and P. J. H. Schoemaker, "Scanning the Periphery," *Harvard Business Review* (November 2005): 135–148.

35. Thomas H. Davenport and Rajeev Ronanki, "Artificial Intelligence for the Real World," *Harvard Business Review* (January/February 2018): 108–116.

36. See "Fast Facts about Vanguard," Vanguard, 1995–2019, https://about.vanguard.com/who-we-are/fast-facts/.

37. Kim Bhasin and Patrick Clark, "How Amazon Triggered a Robot Arms Race," *Bloomberg*, June 29, 2016.

38. Alvin Toffler, *Future Shock* (New York: Bantam Books, 1971), 20. He defined *future shock* as "the distress, both physical and psychological that arises from an overload of the human organism's physical adaptive systems and its decision making process ... the human response to over-simulation."

39. Paul J. H. Schoemaker and Philip E. Tetlock, "Building a More Intelligent Enterprise," *MIT Sloan Management Review* 58, no. 3 (Spring 2017): 28–37.

2 Vigilant Organizations

1. Quoted in Sunil Gupta and Lauren Barley, "Reinventing Adobe," Case 9-514-066, *Harvard Business School,* January 20, 2015, 11.

2. Farhad Manjoo, "Photoshop at 25: A Thriving Chameleon Adapts to an Instagram World," *New York Times*, February 18, 2015.

3. Cited in Gupta and Barley, "Reinventing Adobe," 4.

4. The protests came in response to a petition: Derek Schoffstatt, "Adobe Systems: Eliminate the Mandatory Creative Cloud Subscription," Change.org, May 6, 2013.

5. Gupta and Barley, "Reinventing Adobe," 3.

6. Gupta and Barley, "Reinventing Adobe," 5.

7. Tod Martin, "New Normal Leadership," *Radar: The Journal of Adaptive Change* 1 (May 2016), http://www.radarjournal.com/new-normal-leader.

8. A fuller description of these qualities of vigilant leaders and the research that led us to formulate them can be found in George S. Day and Paul J. H. Schoemaker, "Are You a Vigilant Leader?," *MIT Sloan Management Review* 40 (Spring 2008): 43–51.

9. Paul Michelman, "Key Words for Digital Transformation: Interview of Shantaan Narayen," *MIT Sloan Management Review*, December 4, 2018.

10. W. Glenn Rowe, "Creating Wealth in Organizations: The Role of Strategic Leadership," *Academy of Management Executive* 15 (2001): 81–94.

11. Vara Sprague, "Reborn in the Cloud," *Digital McKinsey*, July 2015, https://www.mckinsey.com/business-functions/digital-mckinsey/our-insights/reborn-in-the-cloud.

12. Gupta and Barley, "Reinventing Adobe," 3.

13. Gupta and Barley, "Reinventing Adobe," 5. The initial reaction was investor skepticism. See Michael Corkery, "Adobe Buys Omniture: What Were They Thinking?," *Wall Street Journal*, September 16, 2009.

14. Gupta and Barley, "Reinventing Adobe," 7.

15. Martin Ihrig, Ian MacMillan, and Jill Steinhour, "Mapping Critical Knowledge for Digital Transformation," *Knowledge@Wharton*, July 6, 2017, https://knowledge.wharton.upenn.edu/article/management-knowledge-assets/.

16. Daniel Lyons, "The Customer Is Always Right," *Newsweek*, January 4, 2010, 85–86.

17. The performance benefits of outside-in approaches have been studied thoroughly by students of *market orientation* (a broader term than *customer-centricity*). For a representative study, see Neil A. Morgan, Douglas Voorhies, and Charlotte Mason, "Market Orientation, Marketing Capabilities, and Firm Performance," *Strategic Management Journal* 30, no. 8 (August 2009): 909–920. For a meta-analysis of many studies that supports a positive and robust relationship between a market orientation and firm performance, see Ahmet H. Kirca, Satish Jayachandran, and William Bearden, "Market Orientation: A Meta-Analytic Review and Assessment of Its Antecedents and Impact on Performance," *Journal of Marketing* 69, no. 2 (April 2005): 24–41.

18. Gupta and Bailey, "Reinventing Adobe," 10.

19. Andrew S. Grove, *Only the Paranoid Survive* (New York: Doubleday, 1996), 115.

3 Managing Organizational Attention

1. Herbert A. Simon, "Designing Organizations for an Information-Rich World," in *Computers, Communication, and the Public Interest*, ed. Martin Greenberger (Baltimore, MD: The Johns Hopkins Press, 1971), 40–41.

2. Dean A. Shepherd, Jeffrey S. McMullen, and William Ocasio, "Is That an Opportunity? An Attention Model of Top Managers' Opportunity Beliefs for Strategic Action," *Strategic Management Journal* 38, no. 3 (2017): 626–644.

3. Margaret Heffernan, *Willful Blindness: Why We Ignore the Obvious* (New York: Simon and Schuster, 2011).

4. Joline Gutierrez Krueger's unfortunate life turn is further explained in Dan Frosch, "Opioids: The Story that Came Home," *Wall Street Journal*, December 9–10, 2017, A1 and A10.

5. Daniel Kahneman, *Attention and Effort* (Englewood Cliffs, NJ: Prentice-Hall, 1973).

6. Henry Mintzberg, *The Nature of Managerial Work* (New York: Harper and Row, 1973).

7. Michael E. Porter and Nitin Nohria, "How CEOs Manage Time," *Harvard Business Review* 96, no. 4 (June/July 2018): 42–51.

8. Day and Schoemaker, *Peripheral Vision.*

9. Christopher Chabris and Daniel Simons, *The Invisible Gorilla: And Other Ways Our Intuitions Deceive Us* (New York: Harmony, 2010), 5–8.

10. John McPhee, *A Sense of Where You Are: Bill Bradley at Princeton* (New York: Farrar, Straus and Giroux, 1999).

11. Jeffrey Ball, "Shell Faces Lower Forever," *Fortune,* February 1, 2018, 84–93.

12. The phrase *radical uncertainty* comes from John Edward King, *A History of Post Keynesian Economics since 1936* (Cheltenham Glos, UK: Edward Elgar Publishing, 2002).

13. Wesley M. Cohen and Daniel A. Levinthal, "Absorptive Capacity: A New Perspective on Learning and Innovation," *Administrative Science Quarterly* (1990): 128–152; and Shaker A. Zahra and Gerard George, "Absorptive Capacity: A Review, Reconceptualization, and Extension," *Academy of Management Review* 27, no. 2 (2002): 185–203.

14. William Ocasio, "Towards an Attention-Based View of the Firm," *Strategic Management Journal* 18, no. S1 (1997): 187–206; William Ocasio, "Attention to Attention," *Organization Science* 22, no. 5 (2011): 1286–1296; and Thomas H. Davenport and John C. Beck, *The Attention Economy: Understanding the New Currency of Business* (Boston: Harvard Business Press, 2001).

15. Andrew S. Grove, *Only the Paranoid Survive: How to Exploit the Crisis Points that Challenge Every Company* (New York: Currency Doubleday, 1999).

16. Jay R. Galbraith, "Organization Design: An Information Processing View," *Interfaces* 4, no. 3 (1974): 28–36.

17. Harbir Singh, "The Relational Organization: From Relational Rents to Alliance Capability," in *The SMS Blackwell Handbook of Organizational Capabilities: Emergence, Development, and Change,* ed. Constance E. Helfat (Malden, MA: Wiley-Blackwell, 2003), 257–263.

18. Teppo Felin and Thomas C. Powell, "Designing Organizations for Dynamic Capabilities," *California Management Review* 58, no. 4 (2016): 78–96; subsequent quotes about Valve were taken from this article.

19. See "Valve Corporation," Wikipedia, https://en.wikipedia.org/wiki/Valve_Corporation#Organizational_structure; also see "Valve," http://www.valvesoftware.com/.

20. Nico Holtzhausen and Jeremias J. de Klerk, "Servant Leadership and the Scrum Team's Effectiveness," *Leadership & Organization Development Journal* 39, no. 7 (2018): 873–882; see also https://www.scrum.org/resources/creators-scrum-ken-schwaber-and-jeff-sutherland-update-scrum-guide-0.

21. T. Lappi, T. Karvonen, L. E. Lwakatare, K. Aaltonen, and P. Kuvaja, "Toward an Improved Understanding of Agile Project Governance: A Systematic Literature Review," *Project Management Journal* 49, no. 6 (2018): 39–63, https://doi.org/10.1177/8756972818803482.

22. Craig Giammona, David Russell, and Brandon Kochkodin, "Slow Pour, Slower Growth," *Bloomberg Businessweek*, April 30, 2018, 15–17.

23. See "Letter: George Shultz Played Key Role in the Rise and Fall of Theranos—It Wasn't Just Elizabeth Holmes," March 25, 2018, https://www.mercurynews.com/2018/03/25/letter-george-shultz-played-key-role-in-the-rise-and-fall-of-theranos-it-wasnt-just-elizabeth-holmes/.

24. John Carreyrou, *Bad Blood: Secrets and Lies in a Silicon Valley Startup* (New York: Knopf, 2018).

25. James O'Toole and David Vogel, "Two and a Half Cheers for Conscious Capitalism," *California Management Review* 53 (Spring 2011): 60–76.

26. Laura J. Keller and Shahien Nasiripour, "Wells Fargo's Uphill Battle," *Bloomberg Businessweek*, March 5, 2018, 31–33.

27. Susan E. Jackson and Jane E. Dutton, "Discerning Threats and Opportunities," *Administrative Science Quarterly* 33 (September 1988): 370–387.

28. See https://www.wired.co.uk/article/what-is-gdpr-uk-eu-legislation-compliance-summary-fines-2018.

29. William P. Smith and Filiz Tabak, "Monitoring Employee E-mails: Is There Any Room for Privacy?," *Academy of Management Perspectives* 23, no. 4 (2009): 33–48; and Kirsten Martin, "Information Technology and Privacy: Conceptual Muddles or Privacy Vacuums?," *Ethics and Information Technology* 14, no. 4 (2012): 267–284.

30. Micah Zenko, *Red Team: How to Succeed by Thinking Like the Enemy* (New York: Basic Books, 2015).

31. William H. Starbuck and Frances J. Milliken, "Challenger: Fine-Tuning the Odds Until Something Breaks," *Journal of Management Studies* 25, no. 4 (1988): 319–340; and Diane Vaughan, *The Challenger Launch Decision: Risky Technology, Culture, and Deviance at NASA* (Chicago: University of Chicago Press, 1997).

4 Sensing Weak Signals Sooner

1. Commonly attributed to Voltaire (1694–1778), which was the pen name for François-Marie Arouet. The original saying was "It is easier to judge the mind of a man by his questions rather than his answers," which was written by Pierre-Marc-Gaston, duc de Lévis (1764–1830), *Maximes et réflexions sur différents sujets de morale et de politique* (Paris, 1808): Maxim xviii; https://en.wikiquote.org/wiki/Voltaire.

2. This diagram was first developed in our earlier book on *Peripheral Vision* to represent the major zones of the periphery and has since been used with our clients to classify weak signals. This particular illustration is a composite from several firms.

3. Eric Schmidt and Joseph Rosenberg, *How Google Works* (New York: Grand Central Publishing, 2014).

4. See Aneela Mirchandani, "The Original Frankenfoods: Origins of Our Fear of Genetic Engineering," *Genetic Literacy Project*, February 10, 2015, https://geneticliteracyproject.org/2015/02/10/the-original-frankenfoods/.

5. William A. Sahlman and Ryland Matthew Willis, "Procter & Gamble 2000 (A): The SpinBrush and Innovation at P&G," Harvard Business School Case 804-099, November 2003.

6. Max H. Bazerman and Michael D. Watkins, *Predictable Surprises: The Disasters You Should Have Seen Coming and How to Prevent Them* (Boston: Harvard Business School Press, 2004).

7. Andrew Grove, *Only the Paranoid Survive* (New York: Currency, 1996).

8. See Patti Strand, "Febreze Is Safe around Pets," National Animal Interest Alliance (NAIA), January 12, 2012, http://www.naiaonline.org/articles/article/febreze-is-safe-around-pets#sthash.3bJ3dvDT.dpbs.

9. Daniel Franklin, *Megatech: Technology in 2050* (London: Economist Books, 2017).

10. Paul J. H. Schoemaker, "Scenario Planning: A Tool for Strategic Thinking," *MIT Sloan Management Review* 36 (Winter 1995): 25–40; and Kees van der Heyden, *Scenarios: The Art of Strategic Conversation* (New York: John Wiley, 1996).

11. P. J. H. Schoemaker and P. Tetlock, "Taboo Scenarios: How to Think about the Unthinkable," *California Management Review* 54, no. 2 (Winter 2012): 5–24.

12. Micah Zenko, *Red Team: How to Succeed by Thinking Like the Enemy* (New York: Basic Books, 2015).

13. Russell L. Ackoff, *Re-creating the Corporation: A Design of Organizations for the 21st Century* (Oxford: Oxford University Press, 1999).

14. Vijay Govindarajan, "Planned Opportunism: Using Weak Signals to Spur Innovation," *Harvard Business Review* 94, no. 5 (May 2016): 54–61.

15. For a good source, see C. J. Nesmeth, B. Persinnaz, M. Persinnaz, and J. A. Goncalo, "The Liberating Role of Conflict in Group Creativity: A Study in Two Countries," *European Journal of Social Psychology* 34 (July/August 2004): 365–374.

16. Roger L. Martin, *The Opposable Mind: Winning through Integrative Thinking* (Boston: Harvard Business Press, 2009).

17. Drake Baer, "Five Brilliant Strategies Jeff Bezos Used to Build the Amazon Empire," *Business Insider*, March 17, 2014.

18. See Paul Schoemaker, "7 Ways to Improve Your Team's Communication," *Inc.*, March 24, 2015, https://www.inc.com/paul-schoemaker/how-to-foster-deep-dialog-in-teams.html.

19. Walter Isaacson, *Steve Jobs* (New York: Simon & Schuster, 2011).

20. James Surowiecki, *The Wisdom of Crowds: Why the Many Are Smarter than the Few and How Collective Wisdom Shapes Business, Economics, Societies and Nations* (New York: Random House, 2005).

21. Paul J. H. Schoemaker, George S. Day, and Scott Snyder, "Integrating Organizational Networks, Weak Signals Strategic Radars and Scenario Planning," *Technological Forecasting and Social Change* 80, no. 4 (May 2013): 815–824.

22. Paul Kleindorfer and Yoram Wind, eds., *The Network Challenge: Strategy, Profit and Risk in an Interlinked World* (Upper Saddle River, NJ: Wharton School Publishing, 2010).

23. Christian Terweisch and Karl Ulrich, *Innovation Tournaments: Creating and Selecting Exceptional Opportunities* (Boston: Harvard Business School Press, 2009).

24. Craig Standing, Denise Jackson, Ann-Claire Larsen, Yuliani Suseno, Richard Fulford, and Denise Gengatharen, "Enhancing Individual Innovation in Organisations: A Review of the Literature," *International Journal of Innovation and Learning* 19, no. 1 (2016): 44–62.

25. George S. Day, "Is It Real? Can We Win? Is It Worth It?," *Harvard Business Review* (December 2007): 3–13.

26. Rita Gunther McGrath and Ian C. MacMillan, *Discovery-Driven Growth: A Breakthrough Process to Reduce Risk and Seize Opportunity* (Boston: Harvard Business Press 2009).

27. This quote, plus others in this paragraph and the next one, come from Kirsten Sandberg, "How Blockchain Could Change Publishing: Part One," *Master of Science in Publishing Blog*, Pace University, February 21, 2018, mspub.blogs.pace.edu/2018/02/21/how-blockchain-could-change-publishing-part-i.

28. See "Using Blockchain: A Strategic Roadmap for Companies," *Knowledge@Wharton*, February 7, 2019, http://knowledge.wharton.upenn.edu/article/using-the-blockchain-a-strategic-roadmap-for-companies/?utm_source=kw_newsletter&utm_medium=email&utm_campaign=2019-02-07.

29. Email from Kirsten Sandberg, March 29, 2019.

30. A more precise analogy would be the World Wide Web (HTTP) and the blockchain as two layers on the Internet stack, each on top of IP/TCP.

31. Philip Tetlock and Dan Gardner, *Superforecasting: The Art and Science of Prediction* (New York: Crown Publishers, 2015); also see Paul J. H. Schoemaker and Philip E. Tetlock, "Superforecasting: How to Upgrade Your Company's Judgment," *Harvard Business Review* (May 2016): 72–78.

32. Robert T. Clemen and Robert L. Winkler, "Unanimity and Compromise among Probability Forecasters," *Management Science* 36, no. 7 (1990): 767–779.

33. Jonathan Baron, Barbara A. Mellers, Philip E. Tetlock, Eric Stone, and Lyle H. Ungar, "Two Reasons to Make Aggregated Probability Forecasts More Extreme," *Decision Analysis* 11, no. 2 (2014): 133–145.

34. Elinor Ostrom, review of *The Difference: How the Power of Diversity Creates Better Groups, Firms, Schools, and Societies*, by Scott E. Page, *Perspectives on Politics* 6, no. 4 (December 2008): 828–829.

35. G. Rowe and G. Wright, eds., "The Delphi Technique: Current Developments in Theory and Practice," special issue, *Technological Forecasting and Social Change* 78, no. 9 (2011).

36. Clarke Claywood, *The Handbook of Strategic Public Relations and Integrated Marketing Communications*, 2nd ed. (New York: McGraw-Hill Professional, 2012).

5 Tackling Ambiguity

1. Quoted in "The Art of Fiction," in *Writers at Work: The Paris Review Interviews, Eighth Series*, ed. George Plimpton (New York: Penguin Books, 1988), 305.

2. Bruce Sterling, "Ten Technologies That Deserve to Die," *MIT Technology Review*, October 1, 2003, https://www.technologyreview.com/s/402039/ten-technologies-that-deserve-to-die/.

3. For a sampling of this diverse array of methods, see Jim Brown, *Change by Design* (New York: HarperCollins, 2009); Eric von Hippel, *Democratizing Innovation* (Cambridge, MA: MIT Press, 2006); Gerald Zaltman, *How Customers Think: Essential Insights into the Mind of the Market* (Boston: Harvard Business School Press, 2003); and Scott Anthony, Mark W. Johnson, Joseph V. Sinfield, and Elizabeth Altman, *The Innovator's Guide to Growth* (Boston: Harvard Business School Press, 2008).

4. Andy Dong, Massimo Garbuio, and Dan Lovallo, "Generative Sensing: A Design Perspective on the Micro-foundations of Sensing Capabilities," *California Management Review* 58, no. 4 (2016): 97–117.

5. Michael Schrage, *The Innovator's Hypothesis: How Cheap Experiments Are Worth More than Good Ideas* (Cambridge, MA: MIT Press, 2014).

6. Mark Kendall and Michael Scholand, *Energy Savings Potential of Solid State Lighting Applications* (Arlington, VA: Arthur D. Little, 2001).

7. See Jaimin S. Patel, "Advanced LED Lighting for Plant Health and Protection in Controlled Environment Agriculture," *Urban Ag News*, September 10, 2016, https://urbanagnews.com/blog/advanced-led-lighting-for-plant-health-and-protection-in-controlled-environment-agriculture/.

8. C. Peirce, *Collected Papers of Charles Sanders Peirce, Vol. I, Principles of Philosophy*, ed. C. Hartshorne and P. Weiss (Cambridge, MA: Harvard University Press, 1932).

9. Abductive reasoning is one of the cognitive underpinnings of the dynamic sensing capability: Dong, Garbuio, and Lovallo, "Generative Sensing." Abduction is also a key element of the process by which a manager forms a belief that

a weak signal represents an opportunity worth pursuing: Dean A. Shepherd, Jeffrey S. McMullen, and William Ocasio, "Is That an Opportunity? An Attention Model of Top Managers' Opportunity Beliefs for Strategic Action," *Strategic Management Journal* 38 (February 2016): 626–644.

10. See Harvey Schachter, "Martin: A New Way of Thinking for a New Way of Business," *Globe and Mail*, October 27, 2011, https://www.theglobeandmail.com/report-on-business/careers/careers-leadership/martin-a-new-way-of-thinking-for-a-new-way-of-business/article559290/.

11. William H. Starbuck, "Organizations as Action Generators," *American Sociological Review* 48, no. 1 (1983): 91–102, https://dx.doi.org/10.2307/2095147. Abduction approximates Popper's (1935) "Logic of Discovery"; see G. Harman, "The Inference to the Best Explanation," *Philosophical Review* 64 (1965): 88–95.

12. See "Lighting New Year's Eve: Ringing in the New Year with Philips LED Lighting," Philips, 2018–2019, https://www.usa.lighting.philips.com/consumer/times-square-ball.

13. Benjamin A. Jones, "Measuring Externalities of Energy Efficiency Investments Using Subjective Well-Being Data: The Case of LED Streetlights," *Resource and Energy Economics* 52 (2018): 18–32.

14. Susan Walsh Sanderson and Kenneth L. Simons, "Light Emitting Diodes and the Lighting Revolution: The Emergence of a Solid-State Lighting Industry," *Research Policy* 43, no. 10 (2014): 1730–1746.

15. See "Philips Takes Control of LED Maker Lumileds," *LEDs Magazine*, August 15, 2005. https://www.ledsmagazine.com/articles/2005/08/philips-takes-control-of-led-maker-lumileds.html.

16. Paul J. H. Schoemaker, "Scenario Planning: A Tool for Strategic Thinking," *Sloan Management Review* (Winter 1995): 25–40.

17. Paul J. H. Schoemaker, "Multiple Scenario Developing: Its Conceptual and Behavioral Basis," *Strategic Management Journal* 14 (1993): 193–213.

18. Govi Rao and Philips Lighting, personal communications.

19. Philips bought Color Kenetics in the United States and TIR in Canada; Color Kenetics especially gave Philips a strong position in the outside lighting of building and city centers with multiple colors and versatile electronic digital control systems.

20. Armatures are the parts of light fixtures that distribute the luminous flux, protect human eyes from bright light rays, deliver the electric current, and reinforce the lamp itself against damage. A luminaire is defined as "a complete lighting unit consisting of a lamp or lamps together with the parts designed to distribute the light, to position and protect the lamps and ballast (where applicable), and to connect the lamps to the power supply"; see Mark C. Ode, "Understanding Luminaires and Lamps," *Electrical Contractor*, June 2003, https://www.ecmag.com/section/codes-standards/understanding-luminaires-and-lamps.

21. Lydie Huché-Thélier et al., "Light Signaling and Plant Responses to Blue and UV Radiations: Perspectives for Applications in Horticulture," *Environmental and Experimental Botany* 121 (2016): 22–38.

22. See Maarten van Tartwijk, "Philips Sells Majority Stake in LED Components, Automotive Business," *Wall Street Journal*, March 31, 2015, https://www.wsj.com/articles/philips-sells-lighting-business-to-consortium-1427780362.

23. See Matt Egan, "GE Can't Get Rid of Its Light Bulb Business," *CNN Business*, May 22, 2018, https://money.cnn.com/2018/05/22/news/companies/general-electric-light-bulb-sale/index.html; GE is still selling its lighting business while profits drop, Philips Lighting has been renamed Signify, and Osram is focused on automotive LED lighting.

24. Stuart Firestein, *Ignorance: How It Drives Science* (Oxford: Oxford University Press, 2012); and Stuart Firestein, *Failure: Why Science Is So Successful* (Oxford: Oxford University Press, 2015).

25. Karen Locke, Karen Golden-Biddle, and Martha S. Feldman, "Making Doubt Generative: Rethinking the Role of Doubt in the Research Process," *Organization Science* 19, no. 6 (2008): 907–918, http://escholarship.org/uc/item/24m701bz.

6 Taking Timely Action

1. From the play *Oedipus Rex* (also called *Oedipus Tyrannus*)—circa 429 BC.; see also *Oedipus the King by Sophocles*, trans. Bernard Knox (New York: Pocket Books, 1959), 35. The quote is said by the character Choral Leader.

2. Rita Gunther McGrath, "Now Hear This: Why the $8 Billion Hearing Aid Sector Is Ripe for Disruption." https://www.ritamcgrath.com/wp-content/uploads/2019/03/March-2018-Newsletter.pdf.

3. Tanguy Catlin and Johannes-Tobias Lorenz, "Digital Disruption in Insurance: Cutting through the Noise," McKinsey & Co., March 2017.

4. Robert J. Crawford, GlaxoSmithKline, U.S. Sales Practices, INSEAD case #06/2015-6024; and Henry A. Waxman, "The lessons of Vioxx," *New England Journal of Medicine* 353 (2005): 1420–1421.

5. This description of the Novartis digital initiative drew from case IMD-32437 by D. A. March and P. Bochukova, "Digital Transformation at Novartis to Improve Customer Engagement," Lausanne, Switzerland, January 1, 2014; as well as from "From Monologue to Dialogue: Fostering Meaningful Engagement with the Medical Community," Novartis AG, 2015; and S. Bennett, "From Snitch Pill to Xbox Sensors, Novartis Goes Digital," *Bloomberg*, March 24, 2015.

6. George S. Day and Paul J. H. Schoemaker, "Managing Uncertainty: Ten Lessons for Green Technologies," *MIT Sloan Management Review* (September 2011): 53–60.

7. For further background on DuPont Biofuels, see George S. Day and Paul J. H. Schoemaker, "Adapting to Fast-Changing Markets and Technologies," *California Management Review* 58 (Summer 2016): 59–77.

8. Avinash K. Dixit and R. S. Pindyck, "The Options Approach to Capital Investment," *Harvard Business Review* (May/June 1995): 105–115; and Ian C. MacMillan and Rita Gunther McGrath, "Crafting R&D Project Portfolios," *Research Technology Management* (September/October 2002): 48–59.

9. Such flexible investments are also called real options or shadow options in contrast to financial ones, which are traded and priced in financial markets. Real options instead reside inside the company and are part and parcel of its on-the-ground business activities and as such would be hard to pull out, trade, or arbitrage externally. Their value is usually assessed using decision trees, NPV analysis, and Monte Carlo simulations perhaps to quantify the benefits of flexibility and the value of future information when viewed ex ante.

10. H. Bahrami and S. Evans, "Super-Flexibility for Real-Time Adaptation: Perspectives from Silicon Valley," *California Management Review* 53, no. 3 (Spring 2011): 21–39.

11. George Stalk Jr. and Ashish Iyer, "How to Hedge Your Strategic Bets," *Harvard Business Review* (May 2016): 81–86.

12. Larry Huston and Nabil Sakkab, "Connect and Develop: Inside Procter & Gamble's New Model for Innovation," *Harvard Business Review* (March 2006):

58–66. This application built on and extends the original conceptualization by Henry Chesbrough, *Open Innovation: The New Imperative for Creating and Profiting from Technology* (Boston: Harvard Business School Press, 2003).

13. Jeffrey H. Dyer and Harbir Singh, "The Relational View: Cooperative Strategy and Sources of Interorganizational Competitive Advantage," *Academy of Management Review* 23 (1998): 660–679.

14. Henry Chesbrough and Marcel Bogers, "Explicating Open Innovation: A New Paradigm for Understanding Industrial Innovation," in *New Frontiers in Open Innovation*, ed. Henry Chesbrough, W. Vandover-beke, and J. West (Oxford: Oxford University Press 2014), 3–28.

15. Andrea Urbinati, David Chiaroni, Vittorio Chiesa, and Frederico Frattine, "The Role of Digital Technologies in Open Innovation Processes: An Exploratory Multiple Case Study Analysis," *R&D Management* (January 2018): 1–25, https://doi.org/10.1111/radm.12313.

16. Marcelo Cano-Kollmann, Snehal Awate, T. J. Hannigan, and Ram Mudambi, "Burying the Hatchet for Catch-Up: Open Innovation among Industry Laggards in the Automotive Industry," *California Management Review* 60 (Winter 2018): 17–42.

17. Jurgen Meffert and Anand Swaminathan, "Management's Next Frontier: Making the Most of the Ecosystem Economy," *Digital McKinsey*, October 2017, https://www.mckinsey.com/business-functions/digital-mckinsey/our-insights/managements-next-frontier.

18. Ron Adner, "Ecosystem as Structure: An Actionable Construct for Strategy," *Journal of Management* 43, no. 1 (January 2017): 39–58.

19. Brent Schendler, "How Big Can Apple Get?," *Fortune* 151, no. 4 (February 21, 2005): 38–45.

20. Kathleen M. Eisenhardt, "Making Fast Strategic Decisions in High-Velocity Environments," *Academy of Management Journal* 32, no. 3 (1989): 543–576; see also Donald Sull and Kathleen M. Eisenhardt, *Simple Rules: How to Thrive in a Complex World* (Boston: Houghton Mifflin Harcourt, 2015).

21. Kathleen M. Eisenhardt and J. Martin, "Dynamic Capabilities: What Are They?," *Strategic Management Journal* 21 (2000): 1105–1121.

22. Jason P. Davis, Kathleen M. Eisenhardt, and Christopher B. Bingham, "Optimal Structure, Market Dynamism, and the Strategy of Simple Rules," *Administrative Science Quarterly* 54, no. 3 (2009): 413–452.

23. K. Eisenhardt and D. Sull, "Strategy as Simple Rules," *Harvard Business Review* 79 (2001): 106–119; and Sull and Eisenhardt, *Simple Rules*.

24. Gary Klein, "Naturalistic Decision Making," *Human Factors: The Journal of the Human Factors and Ergonomics Society* 50, no. 3 (2008): 456–460; and Caroline E. Zsambok and Gary Klein, *Naturalistic Decision Making* (New York: Psychology Press, 2014).

25. D. Kahneman, *Thinking, Fast and Slow* (New York: Farrar, Straus, and Giroux, 2011).

26. Gerd Gigerenzer and P. M. Todd, *Simple Heuristics that Make Us Smart* (Oxford: Oxford University Press, 1999); and Gerd Gigerenzer and Daniel G. Goldstein, "Reasoning the Fast and Frugal Way: Models of Bounded Rationality," *Psychological Review*103, no. 4 (1996): 650.

27. Thomas H. Davenport and George Westerman, "Why So Many High-Profile Digital Transformations Fail," *Harvard Business Review*, March 9, 2018.

28. Ball, "Shell Faces Lower Forever."

29. The basic mechanisms underlying this kind of hype cycle were discussed in our edited book: George S. Day and Paul J. H. Schoemaker, eds., *Wharton on Managing Emerging Technologies* (New York: John Wiley & Sons, 2000). The hype cycle notion developed by the Gartner Group is covered in Jackie Fenn, *Mastering the Hype Cycle: How to Choose the Right Innovation at the Right Time* (Cambridge, MA: Harvard Business School Press, 2018).

30. See Kasey Panetta, "5 Trends Emerge in the Gartner Hype Cycle for Emerging Technologies, 2018," *Smarter with Gartner*, August 16, 2018, https://www.gartner.com/smarterwithgartner/5-trends-emerge-in-gartner-hype-cycle-for-emerging-technologies-2018/.

31. Ozyur Dedhayir and Martin Steinert, "The Hype Cycle Model: A Review and Future Directions," *Technological Forecasting and Social Change* 108 (2016): 28–41.

32. Karl E. Case and Robert J. Shiller, "Is There a Bubble in the Housing Market?," *Brookings Papers on Economic Activity 2003*, no. 2 (2003): 299–342; see also Eugene F. Fame, *The Fama Portfolio: Selected Papers of Eugene F. Fama* (Chicago: University of Chicago Press, 2017).

7 Vigilance

1. "Talking with Intel's Andy Grove," *Forbes*—ASAP Section, February 26, 1996, 64.

2. Among the sources we used to understand the Mastercard transformation described here are Nathan Furr and Andrew Shipilor, "How Does Digital Transformation Happen? The Mastercard Case," INSEAD (Case Center 318-0049-1, 2018); "Mastercard: The Best Kept Platform Secret," Harvard Business School Digital Initiative, March 5, 2018; Avi Salzman, "Why Mastercard Hasn't Been Disrupted," *Barrons*, October 31, 2017; and "Mastercard CEO Ajay Banga's Six Lessons on Leadership," *Quartz India*, April 8, 2018.

3. "MasterCard's Ajay Banga: Why 'Yes If' Is More Powerful than Saying No," *Knowledge@Wharton*, July 24, 2014.

4. "MasterCard's Ajay Banga: Why 'Yes If' Is More Powerful than Saying No."

5. Items in this list are from "MasterCard's Ajay Banga: Why 'Yes If' Is More Powerful than Saying No."

6. Elena Botelho, Kim Powell, Stephen Kincaid, and Dina Wang, "What Sets Successful CEOs Apart," *Harvard Business Review* 95 (May/June 2017): 70–77. The CEO Genome project's research was assembled over ten years and comprises database of assessments of seventeen thousand C-suite executives.

7. "Mastercard CEO Ajay Banga's Six Lessons on Leadership,"

8. See Will Taylor and P. G. LaBarre, *Mavericks at Work: The Most Original Minds in Business Win* (New York: William Morrow, 2006).

9. Francesca Gino, "The Business Case for Curiosity," *Harvard Business Review* 96 (September/October 2018): 48–57.

10. For insights into how narratives provide a generative memory that enables people to translate behavior from the past to inform efforts, see Caroline A. Bartel and Raghu Garud, "The Role of Narratives in Sustaining Organizational Innovation," *Organization Sciences* 20 (January/February 2009): 107–117.

11. Ram Charan, Dennis Carey, and Michael Useem, *Boards That Lead: When to Take Charge, When to Partner, and When to Stay Out of the Way* (Cambridge, MA: Harvard Business Review Press, 2013).

12. There has long been a tension between these two approaches to strategy. On one hand there is the planning school of Michael Porter, starting with his *Competitive Strategy* (New York: Free Press, 1980), and on the other is Henry Mintzberg and the emergent school. See Henry Mintzberg, Bruce Ahlstrand, and Joseph Lampel, *Strategy Safari: A Guided Tour through the Wilds of Strategic Management* (New York: Free Press, 1998).

13. "An anomaly is a fact that doesn't fit received wisdom" is the definition used by Richard P. Rumelt, *Good Strategy/Bad Strategy: The Difference and Why It Matters* (New York: Crown Business, 2011).

14. Geoff Colvin, "How Intuit Reinvents Itself," *Fortune*, November 1, 2017, 77–82.

15. See Donald Sull, "Managing by Commitments," *Harvard Business Review* 81 (June 2003): 53–60. His emphasis is on commitments that restrict a company's future actions. This perspective is also found in Pankaj Ghemawat, *Commitment: The Dynamic of Strategy* (New York: Free Press, 1991), which emphasizes that commitments to developing capabilities superior to those of competitors are often very costly and hard to reverse.

16. Colvin, "How Intuit Reinvents Itself," 82.

17. Laura Heller, "Walmart Launches Tech Incubator Dubbed Store No. 8," *Forbes*, March 20, 2017.

18. Nathan Furr, Kate O'Keefe, and Jeffrey Dyer, "Managing Multiparty Innovation," *Harvard Business Review*, November 2016.

19. See Paul Schoemaker, "How to Rally Your Team around a New Strategy," *Inc.*, June 12, 2012, https://www.inc.com/paul-schoemaker/how-to-get-your-team-onboard-with-a-new-strategy.html; see also chapter 5 in Steven Krupp and Paul J. H. Schoemaker, *Winning the Long Game: How Strategic Leaders Shape the Future* (New York: PublicAffairs, 2014).

20. Chip Heath and Dan Heath, *Switch: How to Change Things When Change Is Hard* (New York: Crown Business, 2010).

21. We are grateful to Michael Taylor, cofounder and principal of Schelling-Point, for his insights into achieving coordinated action. See also Gary Klein, *Seeing What Others Don't: The Remarkable Way We Gain Insights* (New York: PublicAffairs, 2013).

22. Julian Birkinshaw, Alexander Zimmerman, and Sebastian Raisch, "How Do Firms Adapt to Discontinuous Change? Bridging the Dynamic Capabilities and Ambidexterity Perspectives," *California Management Review* 58 (Summer 2016): 36–58.

8 Forewarned Is Forearmed

1. Peter F. Drucker, *Managing in Turbulent Times* (New York: Harper and Row, 1980), 225.

2. Dennis Tourish, "Critical Upward Communication: Ten Commandments for Improving Strategy and Decision Making," *Long Range Planning* 38 (2005): 485–503.

3. Leo Vartorello, "The Lone Creative CEO in Healthcare Wants Company," *Beckers Hospital Review*, June 12, 2018.

4. Sheera Frenkel, Nicholas Confessore, Cecilia Kang, Matthew Rosenberg, and Jack Nicas, "Delay, Deny and Deflect: How Facebook's Leaders Fought Through Crisis," *New York Times*, November 15, 2018, A1, https://www.nytimes.com/2018/11/14/technology/facebook-data-russia-election-racism.html?smprod=nytcore-ipad&smid=nytcore-ipad-share.

5. This is consistent with David Teece, Margaret Peteraf, and Sohvi Leih, "Dynamic Capabilities and Organizational Agility: Risk, Uncertainty and Strategy in the Innovation Economy," *California Management Review* 58, no. 4 (Summer 2016): 13–55; and George S. Day and Paul J. H. Schoemaker, "Adapting to Fast Changing Markets and Technologies," *California Management Review* 58, no. 4 (Summer 2016): 59–77.

6. Thomas Gryta and Ted Mann, "GE: Burned Out," *Wall Street Journal*, December 15, 2018, B14.

7. Roger McNamee, *Zucked: Waking Up to the Facebook Catastrophe* (New York: Penguin Press, 2019).

8. Margaret Heffernan, *Willful Blindness: Why Ignore the Obvious at Your Peril* (New York: Bloomberg, 2011).

9. Clayton M. Christensen, *The Innovators Dilemma: When New Technologies Cause Great Firms to Fail* (Boston: Harvard Business School Press, 1997); and Clayton M. Christensen and Michael E. Raynor, *The Innovator's Solution: Creating and Sustaining Successful Growth* (Boston: Harvard Business School Press, 2003). For a current critique, see Andrew King and Balijr Baatartogtokh, "How Useful Is the Theory of Disruptive Innovation?," *MIT Sloan Management Review* 57, no. 1 (Fall 2015): 76–90; and Christian Hopp, David Antons, Jerman Kaminski, and Torsten Oliver Salge, "The Topic Landscape of Disruption Research: A

Call for Consolidation, Reconciliation and Generalization," *Journal of Product Innovation Management* 35 (2018): 458–487.

10. Charles Darwin, *The Origin of Species* (London: John Murray, 1859).

Appendix B

1. George S. Day and Paul J. H. Schoemaker, *Peripheral Vision: Detecting the Weak Signals That Can Make or Break Your Company* (Cambridge, MA: Harvard Business School Press, 2006).

2. The survey we used to collect data from international companies did not include questions 12A and 12B as shown in appendix A, which both concern the roles of boards (which can vary greatly by industry and country). Also, the endpoints of the scales for A2 and A4 were slightly rephrased in appendix A to make the scoring relative to peers or rivals (which had been the case already in our corporate survey for questions A1 and A3).

3. Christopher Glen Thompson et al., "Extracting the Variance Inflation Factor and Other Multicollinearity Diagnostics from Typical Regression Results," *Basic and Applied Social Psychology* 39, no. 2 (2017): 81–90.

4. Paul J. H. Schoemaker and George S. Day, "Organizational Vigilance: A Comparative Analysis of Three Sectors," submitted to *Futures and Foresight Science*, May 2019 (still under review).

Index

Page numbers followed by f refer to figures.